T0176161

CORROSION FAILURES

WILEY SERIES IN CORROSION

R. Winston Revie, Series Editor

CORROSION FAILURES

Theory, Case Studies, and Solutions

K. ELAYAPERUMAL

Corrosion and Metallurgical Consultant, 3034, Sobha Amethyst, Kannamangala, Bangalore 560 067 India

V.S. RAJA

Professor, Dept. of Metallurgical Engineering and Materials Science,
Indian Institute of Technology Bombay, Powai, Mumbai, India

Published by John Wiley & Sons, Inc., Hoboken, New Jersey
Published simultaneously in Canada

For general information on our other products and services or for technical support, please contact our Customer Care Department within the United States at (800) 762-2974, outside the United States at (317) 572-3993 or fax (317) 572-4002.

Wiley also publishes its books in a variety of electronic formats. Some content that appears in print may not be available in electronic formats. For more information about Wiley products, visit our web site at www.wiley.com.

Library of Congress Cataloging-in-Publication Data:

Elayaperumal, K. (Kailasanathan), 1937- Raja,V.S (Vngaranahalli Srinivasan),1958-
  Corrosion failures: theory, case studies, and solutions/K. Elayaperumal, V.S. Raja.
    pages cm
  Includes bibliographical references and index.
  ISBN 978-0-470-45564-7 (cloth)
  1. Corrosion and anti-corrosives. 2. Metals- Stress corrosion cracking. I. Title.
  TA418.74.E42 2015
  620.1'1223–dc23

                                                                    2015000046

Cover image courtesy of V.S. Raja

Typeset in 11pt/13pt Times by Laserwords Private Limited, Chennai, India

Printed in the United States of America

10 9 8 7 6 5 4 3 2 1

1 2015

CONTENTS

ABOUT THE AUTHORS

Dr. K. Elayaperumal holds degrees in Metallurgical Engineering from Indian Institute of Science, Bangalore, India, and Massachusetts Institute of Technology, Cambridge, Massachusetts, USA. He has acquired vast R&D experience in the field of corrosion of metallic materials in the nuclear power industry and the related chemical process industry in his two-decade career in the Department of Atomic Energy, Government of India, Bhabha Atomic Research Centre. His advisory consultancy services in the field of analysis of corrosion failure in chemical process industries have generated a great amount of interesting case studies and brought out in this book. He is a recipient of National Metallurgist Award by Govt. of India and Life Time Achievement Award by NACE India Section.

Prof. V. S. Raja received his doctoral degree from Indian Institute of Science, Bangalore, India. He is a Professor of Corrosion Science and Engineering in the Department of Metallurgical Engineering and Materials Science and also Institute Chair Professor in the Indian Institute of Technology Bombay, India. For the past 28 years, he taught several corrosion and materials related courses, supervising about 20 doctoral students. His main research lies in understanding the interrelation between metallurgy and corrosion, especially passivity, localized corrosion, and stress corrosion cracking. He is also actively involved in solving industrial problems related to corrosion. He has received several awards for his contribution to teaching, research, and industry and is a NACE International Fellow.

FOREWORD

In this unique book, the authors provide a concise presentation of the essence of corrosion principles with an orientation toward corrosion failures and management of corrosion to prevent failures. By appropriate design, construction, operation, and maintenance, the cost of failure can, indeed, be avoided. The book is divided into the following six chapters that are essential reading for any person concerned with the use of materials in applications where safety, integrity, and reliable, cost-effective operation are required:

1. Introduction: The significance of corrosion failures is described in economic and engineering terms.
2. Thermodynamics and kinetics of corrosion: The basic principles of corrosion are presented along with an outline of techniques for assessing the extent of corrosion in engineering installations.
3. Forms of corrosion: Ten basic forms of corrosion are described, including the critical controlling factors for each form. The authors present a valuable collection of photographs to illustrate the forms of corrosion.
4. Materials of construction for chemical process industries: An extensive discussion of the materials available is presented, along with properties on which materials selection is based. The discussion includes not only metallic materials but also nonmetals – ceramics, glasses, and polymers.
5. Failure analysis procedures with reference to corrosion failures: The steps involved in failure analysis are succinctly described from the

perspective of developing an understanding of the cause of failure as necessary to prevent future failures.

6. Case studies: Eighty case studies are photographically illustrated, presented, and discussed. Each case study includes a description of the type of service involved, the specific problem, the material, observations, diagnosis, and remedy.

I am delighted to commend this book to all readers with an interest in reliable application of engineering materials, and I congratulate the authors on their success in achieving the completion of this valuable book.

R. Winston Revie
CANMET Materials Technology Laboratory
Ottawa, Ontario
Canada

PREFACE

Corrosion cuts across almost all industries, and the corroded components prematurely fail and/or operate at suboptimal level. It affects productivity, safety, and environments and devours nonrenewable natural resources and therefore is of great concern to society. Although industrial components do fail by other mechanisms such as mechanical overload, fatigue, creep, brittle fracture, corrosion remains the single most destruction causing the highest damage to all industrial components. For example, in chemical processing industries, corrosion of process equipment vessels and the associated piping, tubing, and utility components such as boilers, heat exchangers, and condensers is the most predominant damage-causing mechanism among the others mentioned earlier. Corrosion takes place on the inside surface of the vessels and other components because of the action of process chemicals and also on the outside surfaces because of the atmospheric moisture laden with chemical vapors and also by wet insulation materials. On the other hand, corrosion damage in aerospace and other transportation industry is relatively less, though found to be critical in life and safety and in the cost of the components. It is estimated that corrosion causes loss of about 3.5% of GNP of a nation. Hence, corrosion control becomes an important subject of science and engineering.

Corrosion is a multidisciplinary subject that calls for greater/concurrent understanding of electrochemical concepts, materials science, especially metallurgy and design. Teaching corrosion requires not only a sound understanding of these subjects but also better appreciation of actual corrosion problems that industries encounter. On the other hand, industrial corrosion problems are typically handled by mechanical engineers, chemical engineers, and at times metallurgical engineers. What makes solving corrosion problems on the field difficult is that environmental and operating conditions are usually diverse and there is no list of do's and don'ts and nor there exist simple mathematical equations that could correlate the failures to these conditions that alone let the prediction of corrosion failures. In a vast majority of industries, baring a few nuclear power plants and pipelines carrying crude and petrochemicals where the environments are reasonably defined, controlling corrosion is primarily phenomenological in nature. Thus, effective corrosion control by these field engineers can be possible only if they can understand the complex concepts of electrochemistry and metallurgy that govern the corrosion processes. The main objectives of this book are twofold: enable the students to appreciate how the concepts of electrochemistry and metallurgy are intimately related to corrosion failures and empower the practicing engineers (involved in design and manufacturing of industrial components and those involved in process control, inspection, and maintenance) to tackle corrosion problems through simplifying electrochemical concepts and corrosion mechanisms and give exposure to metallurgy and failure analysis methodology and the relevant tools needed for the analysis. The book therefore can serve as text books as much as a reference one. In this sense, this book is considered unique and different from the normal text and failure analysis books published in the subject of corrosion.

The book brings out the phenomenon, importance, and cost of corrosion in various industrial sectors and infrastructures in the first chapter in order to emphasize the need to seriously consider corrosion control. Corrosion processes start with electron transfer from metallic surface to chemical species of the environment, which forms the basis for all types of corrosion. The governing electrochemical thermodynamics decides if a metal can corrode at all in a given environment, while the corrosion rates are decided by electrochemical kinetics. These concepts and the governing equations are simplified and presented in Chapter 2. Of particular importance is the role of polarization (overvoltage), passivity, and their dependence on environment and its relation to corrosion rate and electrochemical corrosion testing and monitoring techniques. The reader is expected to get clarity on these aspects of corrosion.

Chapter 3 covers the different manifestations (can also be called mechanisms) of corrosion that arise out of complexities in environments and metallic structures. For example, corrosive environment can be quiescent or under flow in relation to the metallic structures and it may contain chlorides, foulants, and microbes. On the other hand, the metallic structures could have been fabricated through joining methods such as flange/riveting/welding, they could be under applied stresses or residual stresses, and could be bimetallic in nature. Such situations lead to different forms of corrosion, such as pitting corrosion, crevice corrosion, galvanic corrosion, intergranular corrosion (IGC), flow-assisted corrosion, and environmentally assisted cracking. This chapter is expected to give strong basis for implementing various corrosion control measures such as materials selection as well as for corrosion failure analysis.

Among the corrosion preventive methods such as industrial painting, adding corrosion inhibitors, cathodic protection, and materials selection, the most often considered method for chemical process equipments demanding corrosion resistance and in many cases heat transfer as well is the materials selection. The Chapter 4 deals with this aspect in great detail, starting from plain carbon steels through stainless steels, copper-based alloys, and nickel-based alloys to metals such as titanium and tantalum in the order of increasing corrosion resistance and the accompanying increase in unit cost. To give a broad perspective, light alloy such as aluminum and nonmetallic materials similar to engineering plastics such as polytetrafluoroethylene, fiber-reinforced plastics, ceramics, and glasses are covered.

As mentioned earlier in this preface, detailed analysis of every unexpected corrosion failure is very important and a must to arrive at the root cause of the failure and to take appropriate preventive measures. The process namely failure analysis needs to be done in a systematic scientific way. Chapter 5 deals with systematic scientific procedures of conducting failure analysis starting with field inspection through sample testing and analysis to report writing.

Chapter 6, the longest chapter in the book, presents a set of 80 real-time case studies of actual corrosion failures that occurred in chemical process industries in the very recent past and analyzed by the authors. For each case, details such as the industry group, the equipment identification, the experienced corrosion phenomenon, material of construction (MOC), chemical service condition, tests carried out, diagnostic analysis, and finally recommendations for remedial actions are provided. The needed explanation such as description/mechanisms/characteristics of the particular form of corrosion in question, detailed composition of the MOC used, if needed, can be found in the preceding chapters, that is, Chapters 2–5. This chapter also gives its own content list in which the reader can choose a specific case study of his/her interest and read it in detail.

In summary, the book is an attempt to spread in detail the practical aspects of corrosion failures of chemical process equipments supported by scientific background and explanation of each corrosion phenomenon experienced and reported in the case studies. One can notice a good number of cases for most prominent forms of corrosion occurred in a variety of industries.

It is hoped that the book would serve as a useful practical guide for practicing engineers in their endeavor to control corrosion and also see theory in each and every corrosion failures they encounter, and an inspiring textbook for undergraduate/masters students to see the immediate application of theory in practice. In essence, the authors will be happy if this book can synergize theory and practice. The authors express many thanks to Dr. Winston Revie for encouraging us to write this book as well for writing a foreword and to E. Shanmuganathan, Naresh Ingle and Farooq Mohamed Khan for their assistance in preparing this book and M/s John Wiley & Sons, Inc., for publishing this book.

K. Elayaperumal
V. S. Raja

1

INTRODUCTION

1.1 THE PHENOMENON OF CORROSION

The term rusting has been in vogue long before the human kind initiated any systematic study on corrosion of metals. This term, however, refers to uniform corrosion of steels. If the metal is carbon steel and if the environment is simple humid air, the former corrodes giving "rust" as the final corrosion product, which is seen as a brownish crust/porous scale over the steel surface. The result of such a corrosion phenomenon is the general uniform loss of thickness of the metal and this type of corrosion is generally called "Uniform Corrosion" and is the most common form of corrosion accounting for a major percentage of overall metal losses. Unfortunately, the phenomenon of corrosion is spontaneous in nature supported by thermodynamics. That is to say corrosion lowers the energy of metals, ironically supplied by the mankind to produce metal from their respective ores, to transform to their natural lower energy states such as oxides, sulfides, chlorides, etc.

The environments that give rise to corrosion of metals vary from mildest humid air atmosphere which we all breath-in daily to the most aggressive highly acidic solutions and high temperature gases in which processes such as chlorination, sulfidation, etc. take place. While the mild atmospheres occur mostly under the domestic conditions, including marine atmospheres affecting coastal structures, the severe atmospheres occur among industrial

Corrosion Failures: Theory, Case Studies, and Solutions, First Edition. K. Elayaperumal and V. S. Raja.
© 2015 John Wiley & Sons, Inc. Published 2015 by John Wiley & Sons, Inc.

processes like pickling of metals, chemical processes, power generation, oil and gas production, electronic component processing, transportation industries, etc. Among the industrial processes, the size of the metallic components involved varies from microscopic as in electronics industries to very macroscopic, such as storage and pressure vessels, cross-country piping, heat exchangers, etc.

"Corrosion Failure" is the ultimate result of corrosion. The component, structure, or equipment loses its functionality as a result of corrosion leading to grave consequences. Ultimate failure due to corrosion occurs because, among several reasons, the phenomenon of corrosion has been occurring unabated over a long period without a warning signal. Corrosion phenomenon is in general, not always, a time bound phenomenon. Failures of structures and components due to corrosion in mild atmospheres such as humid air and marine atmospheres would be of minor consequences such as premature replacement cost, temporary public discomfort, etc. On the other hand, unexpected corrosion failures of equipments in chemical process equipments, which the present book addresses, would result in major consequences such as possible leakage of corrosive fluids/vapors, very expensive replacement of equipments, heavy production losses, and at times human fatalities also.

1.2   IMPORTANCE OF CORROSION

Before giving the basics and the case studies, an attempt is made in the following section of this chapter to briefly present the existing information available on the overall costs of corrosion affecting an industrialized nation, both direct and indirect, particularly with respect to chemical process industries

1.2.1   Cost of Corrosion: Direct and Indirect

Metallic corrosion is a major loss-producing phenomenon in many sectors of a nation's economy. This is because corrosion results in loss of metals and materials, energy, labor, etc., which would have been contributively productive otherwise for some other useful purpose. Revie and Uhlig (2008) divide the losses due to corrosion into two categories:

- Direct loss and
- Indirect loss.

Direct losses include:

- Cost of replacing corroded/failed structures/equipments/components,
- Painting and re-painting of corrosion-prone structures to prevent general atmospheric corrosion,

- Costs involved in all other protective measures, such as cathodic protection, inhibitor addition, protective coating/wrapping/cladding, galvanizing, electroplating, etc.,
- Extra cost involved in choosing corrosion-resistant alloys (CRAs) such as stainless steels, nickel base alloys, titanium, etc. in the place of carbon steels which would have been otherwise suitable from mechanical/structural points of view, and
- Cost of dehumidifying storage rooms for storing metallic components/equipments and spare parts, etc. before they are put into use.

Indirect losses are like consequential losses that add heavily, many times very heavily, to the direct losses outlined above. These indirect losses include:

- Loss-of-Production (Downtime) Cost: This factor alone, many times, is orders of magnitude higher than the direct replacement cost,
- Product loss through leaks/failures due to corrosion: This also would be very heavy if the equipment is concerned is a pressure vessel and high pressure pipeline carrying huge quantities of finished products under pressure like utility gas separated from oil, purified potable water through water mains, high pressure steam, etc.,
- Loss of efficiency in heat transfer equipments and pipelines: Accumulation of corrosion product scales on pipelines and on heat transfer surfaces reduces the pumping and heat transfer efficiency, respectively, thereby necessitating increased power to the pumps and heat exchangers,
- Contamination and hence rejection of product: Heavy metal impurities as a result of corrosion of the container equipments and transfer pipelines would result in total rejection of several batches (huge quantities) of the carefully produced (value added) chemical product,
- Over-design: Giving "corrosion allowance," thereby using vessels with thickness much greater than that demanded by mechanical requirements amounts to over-design and adds up to huge indirect cost involved in providing excess metal for corrosion to take place.

The above direct and indirect losses are somewhat quantifiable. But loss of life due to leakage, corrosion fracture, explosion and similar unpredictable corrosion-related failures and accidents cannot be easily quantified but would result in huge compensation losses.

As far as corrosion costs in terms of money values are concerned, the most often quoted estimate is that of the 1998 US Study jointly carried out by US Department of Transportation, and the NACE, the results of which were first published in 2002 (Koch et al. 2002). As per this study report, corrosion

TABLE 1.1   Summary of Industry Sector Direct Corrosion Costs Analyzed in 1998 US Study (with Permission from Federal Highway Administration, USA)

	Estimated Direct Cost of Corrosion Per Category	
Category	$x Billion	%
Infrastructure (highway bridges, gas & liquid transmission pipelines, waterways and ports, hazardous materials storage, airports, railroads)	$22.6	16.4
Utilities (gas distribution, drinking water and sewer systems, electrical utilities, telecommunications)	$47.9	34.7
Transportation (motor vehicles, ships, aircraft, railroad cars, hazardous materials transport)	$29.7	21.5
Production and manufacturing (oil and gas exploration and production, mining, petroleum refining, chemical, petrochemical and pharmaceutical, pulp and paper, agricultural, food processing, electronics, home appliances)	$17.6	12.8
Government (defense, nuclear waste storage)	$20.1	14.6
Total	$137.9	100

losses suffered by Industry and by Government (Total Economy) amount to many billions of dollars annually, approximately US $276 billion in USA alone, about 3.1% of Gross Domestic Product. Out of these, loss in Industries alone amounted to $138 billion annually, as shown in the following break-ups extracted from the above study, Table 1.1.

The figures corresponding to Production and Manufacturing from Table 1.1 amounting to $17.6 billion are shown in the break-up pie-chart form in Figure 1.1 of this chapter, again extracted from the reference Koch et al. (2002).

The costs shown in the above illustrations are direct costs only. The figures do not include indirect costs of production outages resulting from unexpected failures, quite common in chemical process industries. Also the figures do not include those of operation and maintenance related to corrosion only. This is an annual recurring expenditure.

One can notice that for Production and Manufacturing alone, mostly varieties of chemical processing, with which the present book is concerned, the sub total cost is $17.6 billion, 12.8% of total Industry Cost of $137.9 billion. This is an enormous figure by any standard and every attempt should be continuously made to reduce/prevent this great loss due to corrosion.

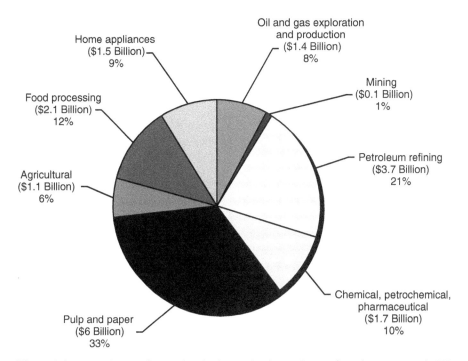

Figure 1.1   Annual cost of corrosion in the production and manufacturing category in US ($17.6 Billion). Koch et al. (2002) (courtesy Federal Highway Administration, USA.)

Among the overall corrosion prevention management strategies suggested include the following Koch et al. (2002):

- Dissemination of corrosion awareness information generated in various industry sectors.
- Increase awareness of significant corrosion costs and potential cost savings.
- Improve education and training of personnel in recognition of corrosion damage and its control.
- Implement advance design practices for better corrosion management.
- Improve corrosion technology through research, development, and implementation.

The present book, giving real-life case histories of major corrosion failures, is aimed at contributing to the above corrosion prevention strategies to varying extent. The case studies originate from various industry sectors and the knowledge gained in analysis is disseminated through the book to the overall corrosion-conscious industry/research personnel and graduate students in general. The book is presented in a simple practical way, without going in depth to advanced scientific and theoretical aspects, so that there is increase in corrosion awareness among the plant personnel and there is improvement in training in corrosion control. Further, the case studies contain remedial measures relating to design improvements, to improved corrosion prevention technologies such as use of CRAs, and to use of improved inspection techniques.

The book would have fulfilled its purpose if it helps to avoid corrosion failures, similar to those presented in the case studies, which are quite likely to happen in the chemical process industries throughout the world.

1.3 THE PURPOSE AND FORMAT OF THE BOOK

The present book addresses itself to the latter failures mentioned above, namely those occurring in chemical process industries. Each unexpected failure deserves a detailed failure analysis study. Due to the social pressure of maintaining production levels of various essential products such as electric power, drinking water, chemicals for human needs, etc., immediate replacement of the failed component/equipment is of utmost priority. But, just the replacement alone would not give assurance/confidence against reoccurrence of similar failures in future in the replaced fresh equipment. Towards this assurance, a detailed scientific analysis/study of the occurred failure with the purpose of arriving at the root cause of the failure (corrosion or otherwise, type of corrosion, what caused it, etc.) should be carried out within a short period from the failure incident date.

The authors of the present book have carried out several such analyses in the past two or three decades and are presenting them in this book in an organized systematic manner in the form of case studies. Dissemination of information/experience gained in each of the failure analysis study thus conducted is the primary purpose of the book.

There are several books and handbooks that either deal with "Corrosion" or "Failure Analysis." But there seems to be no book devoted exclusively to deal with Case Histories related to Corrosion Failures in Chemical Process Industries. Hence, we felt a need for book devoted to this subject. However, understanding any technical phenomenon occurring in any industry requires some familiarity with the scientific basis of the phenomenon. This is true for corrosion failures also. With this latter requirement in mind, the book devotes a few chapters to Basics of Corrosion and Corrosion Monitoring, Forms of Corrosion, Materials of Construction, and Failure Analysis Procedures in a non-elaborative way before describing the case histories in detail. These initial chapters provide the necessary basics to understand the case-studies presented in the last Chapter. So, the Case Studies and the related basics are given in the same book. Such a combination is not available presently.

REFERENCES

Koch, H. G., Brongers, M. P. H., Thompson, N. G., Virmani, Y. P. and Payer, J. H., (2002), Corrosion Costs and Preventive Strategies in the United States. Publication No. FHWA-RD-01-156, Federal Highway Administration, USA.

Revie, W. R. and Uhlig, H. H., (2008), *Corrosion and Corrosion Control: An Introduction to Corrosion Science and Engineering*, 4th Edition, Wiley Interscience, Chapter 1, pp. 02–04.

2

THERMODYNAMICS AND KINETICS OF ELECTROCHEMICAL CORROSION

2.1   INTRODUCTION

Industrial corrosion failures are largely complex in nature. They concern with issues such as suitability of material of construction for a given application, on one hand, and shortcoming in fabrication and commissioning of plants, operational, and maintenance issues at the plants on the other. With such complex issues contributing to corrosion failures, an attempt to correlate corrosion failures to thermodynamics and kinetics of corrosion processes may seem superfluous. Nevertheless, a broad knowledge on this subject is desirable as it can enable an engineer to have an idea not only of the tendency of metals and alloys to corrode but also of the rate at which they corrode, provided a prior knowledge of the environmental conditions is available. Keeping this in mind, this chapter intends to bring out the basic concepts without a rigorous treatment. More detailed treatment of this subject can be found in other books (Bockris and Reddy 1977; Revie and Uhlig 2008; Fontana 1986).

Corrosion Failures: Theory, Case Studies, and Solutions, First Edition. K. Elayaperumal and V. S. Raja.
© 2015 John Wiley & Sons, Inc. Published 2015 by John Wiley & Sons, Inc.

2.2 THERMODYNAMICS

2.2.1 Corrosion Reactions and Gibbs Free Energy Change

At the outset it is necessary to pose the question, "why do metals/alloys corrode?" and try to find a simplified answer that is universal in nature. As discussed in the earlier chapter, the corrosion of metals and alloys involves oxidation from their metallic state and therefore must obey the thermodynamic criteria. Barring a few metals, such as gold and platinum, much of them are found in nature as ores (oxide, sulfides, etc.). So, considerable amount of energy is expended to convert these ores into respective metals. As a result, they remain at higher energy levels than their corresponding ores. Therefore, it is not surprising that most of these metals tend to go back to their low energy state (oxides, chlorides, sulfates, etc.) on exposure to chemical environments. Taking iron—a metal commonly used—as an example the above concept is illustrated in Figure 2.1.

It is interesting to note that corrosion, as a process, releases energy and so is spontaneous in nature. This indeed is a serious problem for almost all the engineering metals and alloys. Applying the Gibbs free energy concept, all the corrosion processes must obey the following relation

$$\Delta G = -\text{ve (negative)} \qquad (2.1)$$

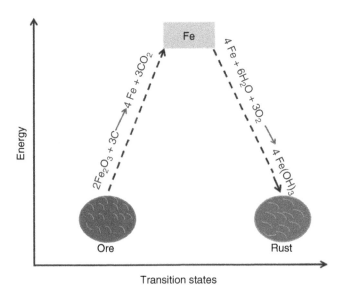

Figure 2.1 Illustration of physical meaning of corrosion and the associated energy is brought out in the figure. Note that energy is supplied to iron ore to convert it into iron and when the latter rusts energy is released.

The free energy change in turn is governed by change in enthalpy (ΔH) and entropy (ΔS) associated with any corrosion reaction. This is given by the relation

$$\Delta G = \Delta H - T\Delta S \qquad (2.2)$$

While the Gibbs free energy concept to predict corrosion of any metal seems simple and appealing, direct measurement of free energy change of any corrosive system does not appear to be quite simple. Hence, there must be better means of predicting the corrosion tendency of a metallic system. Electrochemical concept comes in handy. This aspect is discussed below.

2.2.2 Electrochemical Nature of Corrosion

Corrosion of iron in an acid, say hydrochloric, is a simple "chemical" reaction that can be used to understand how electrochemical concepts can be applied to predict corrosion tendency of a metal.

$$Fe + 2HCl \rightarrow FeCl_2 + H_2 \qquad (2.3)$$

Though the end products are ferrous chloride ($FeCl_2$) and H_2, a closer look at the above equation reveals the fact that iron is oxidized to Fe^{2+} and H^+ is reduced to H_2 as the corrosion of iron proceeds.[1] In other words, the following oxidation and reduction reactions occur when iron is immersed in hydrochloric acid

$$Fe \rightarrow Fe^{2+} + 2e \text{ (oxidation)} \qquad (2.4)$$

$$2H^+ + e \rightarrow H_2 \text{ (reduction)} \qquad (2.5)$$

As these two reactions occur on metallic (Fe) surface, the latter acts as conduit for transfer of electrons from the oxidation site to reduction site (Fig. 2.2). As a result, these reactions become electrochemical in nature. Accordingly, electrochemical criteria can be applied to predict the conditions at which the above two reactions can occur spontaneously.

2.2.2.1 Relation Between Gibbs Free Energy Change and Electrochemical Potential Before we get into establishing electrochemical criteria for a spontaneous reaction, it is essential to look at the relation between the free energy change (ΔG) and the electrochemical potential (E). The relation between free energy change and the so-called electrochemical potential was

---

[1]$FeCl_2$ can be written as $Fe^{2+}2Cl^-$ and HCl can be written as H^+ and Cl^-. The species Cl^- remains unchanged during the corrosion process.

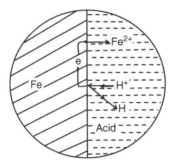

Figure 2.2 A magnified view of the corrosion process occurring on a iron surface exposed to acid solution.

founded by the great scientist, Walter Nernst. According to him they bear the following relation.

$$\Delta G = -nFE \tag{2.6}$$

In the above relation, n, F, and E, respectively, correspond to the number of electrons involved in the electrochemical reaction, Faraday constant ($\sim$96,500 coulombs), and electrochemical potential. As ΔG bears the following relation to standard free energy change (ΔG°) and equilibrium constant[2] (K)

$$\Delta G = \Delta G^\circ + \text{RT} \ln K \tag{2.7}$$

The Equation (2.6) can be written as

$$E = E^\circ - \frac{\text{RT}}{nF} \ln K \tag{2.8}$$

Thus, an electrochemical reaction can be described using standard potential (E°) and equilibrium constant, which are easily measurable parameters. The above equation is called Nernst Equation.

The above equation forms the basis of prediction of corrosion tendency of a metal. It is worthwhile to clarify on the appropriate use of the above equation.

2.2.2.2 Use of Nernst Equation   The best way to examine the application of Nernst equation to an electrochemical reaction is to apply this to the example we have so far considered, namely, the oxidation of iron. One can start with an (electrochemical) equilibrium process involving Fe and Fe^{2+} (consider a piece of metallic iron immersed in an aqueous solution having

---

[2] $K = \frac{a_{\text{product}}}{a_{\text{reactant}}}$. By default only activity of metallic ions is considered. For simplicity and for dilute solutions activity of species can be equated to concentration.

Fe^{2+} ions) and then examine what prompts the equilibrium to drift towards oxidation. This equilibrium can be represented as below.[3]

$$Fe^{2+} + 2e = Fe \qquad (2.9)$$

The corresponding Nernst equation is as follows

$$E = E^\circ - \frac{RT}{2F} \ln \frac{a_{Fe}}{a_{Fe^{2+}}} \qquad (2.10)$$

Looking at Equation (2.10), it can be said that the standard potential equals equilibrium potential, should the activity of Fe (product in any equilibrium) and Fe^{2+} (reactants in any equilibrium) becomes unity.[4] Such an equilibrium potential is called standard potential. Values of standard potential can be summarized either in ascending or in descending values such as those shown in Table 2.1. Such a series is called electro motive force (EMF) series. The following characteristics of the EMF series are important to understand the corrosion tendency of any metal.

(a) Hydrogen equilibrium bears zero value. This value, in fact, is not measured but assigned to this equilibrium.
(b) The other potentials are measured with respect to standard hydrogen electrode (SHE) and so are objective and sign invariant.[5]
(c) Those equilibria having higher standard potential than that of H$^+$/H will be noble while the others having lower standard potential will be active and corrode in acid solutions.

The points which need to be emphasized here are that the equilibrium potential (E) depends not only on the standard potential (E°) (which is unique to any electrochemical equilibrium), but also on the concentrations of the reactants and products. Therefore, the tendency of a metal to corrode can be altered irrespective of its inherent tendency to corrode. This aspect will be further elaborated considering a more general case, than standard states described here.

2.2.2.3 Electrochemical Basis for Prediction Having suggested that electrochemical potential forms basis for predicting corrosion tendency of a metal, it is necessary to formulate unambiguous guidelines (i) to determine

---

[3] As per IUPAC convention, the equilibrium is written with oxidized species as reactants.
[4] Activity of pure metal/ element is considered unity.
[5] Earlier American system employed the so-called oxidation potential and the European the so-called reduction potential.

TABLE 2.1   Standard Potential of Various Electrochemical Equilibria at 25 °C (77 °F), Volts versus SHE

$Au^{3+} + 3e^- = Au$	+1.498
$O_2 + 4H^+ + 4e^- = 2H_2O$	+1.229
$Pt^{2+} + 2e^- = Pt$	+1.20
$Pd^{2+} + 2e^- = Pd$	+0.987
$Ag^+ + e^- = Ag$	+0.800
$Hg^{2+} + 2e^- = Hg$	+0.854
$Hg^{2+} + 2e^- = 2Hg$	+0.789
$Fe^{3+} + e^- = Fe^{2+}$	+0.771
$O_2 + 2H_2O + 4e^- = 4OH$	+0.401
$Cu^{2+} + 2e^- = Cu$	+0.337
$2H^+ + 2e^- = H_2$	0.000 (Reference)
$Pb^{2+} + 2e^- = Pb$	−0.126
$Sn^{2+} + 2e^- = Sn$	−0.136
$Ni^{2+} + 2e^- = Ni$	−0.250
$Co^{2+} + 2e^- = Ni$	−0.277
$Tl^+ + e^- = Tl$	−0.336
$In^{3+} + 3e^- = In$	−0.342
$Cd^{2+} + 2e^- = Cd$	−0.403
$Fe^{2+} + 2e^- = Fe$	−0.440
$Cr^{3+} + 3e^- = Cr$	−0.744
$Zn^{2+} + 2e^- = Zn$	−0.763
$Al^{3+} + 3e^- = Al$	−1.662
$Mg^{2+} + 2e^- = Mg$	−2.363
$Na^+ + e^- = Na$	−2.714
$K^+ + e^- = K$	−2.925

whether a metal can corrode under a given condition and (ii) to suggest when a metal can acquire a tendency to corrode. For doing this, one can start with examining an electrochemical cell. In any electrochemical cell there would be anodic (oxidation) and cathodic (reduction) reactions which will give rise to a positive cell voltage (E_{cell}). Actually, an electrochemical cell starts with at least two independent electrochemical equilibria (also called half-cells). On coupling one of them turns into anodic and the other cathodic such that the cell potential becomes positive. E_{cell} is given by the relation[6]

$$E_{cell} = E_c - E_a \qquad (2.11)$$

[6]Some authors employ short cuts using combination of oxidation and reduction potentials to simplify the problem. While mathematically it is correct, it lacks objectivity and is against the well established norms that equilibrium potentials are sign invariant.

E_c and E_a correspond to equilibrium potentials of the "so-called" cathodic and anodic reactions (half-cells). Note that E_c and E_a are not standard potentials, but are equilibrium potentials obtained as per Nernst equation given in Equations (2.8) and (2.10).

Let us consider two cases, namely Cu and Zn, and predict if they could corrode in deaerated acid. For simplicity, Cu, Zn, and H^+/H (acid) equilibria are considered in their standard states (otherwise, if activities of Cu^{2+}, Fe^{2+}, and H^+ and fugacity of H_2 are known, E can be determined based on Nernst equation). From the table of standard potentials, those of Cu^{2+}/Cu, Zn^{2+}/Zn, and H^+/H are found to be respectively +0.337, −0.763, and 0.0 V. In case of Cu^{2+}/Cu and H^+/H equilibria, an electrochemical cell with a positive potential is possible, only if the former turns into cathodic and so copper will not corrode. On the other hand, when Zn^{2+}/Zn and H^+/H equilibria are allowed to couple, the resulting electrochemical cell will give rise to a positive potential only when Zn^{2+}/Zn equilibrium turns into anodic and H^+/H cathodic. Accordingly, Zn in acid suffers corrosion. Similar exercise can be done with other possible equilibriums, such as O_2/OH and O_2/H_2O, to examine how they are different from acid solutions.

2.2.2.4 Stability Diagrams By comparing the equilibrium potentials (half cell potential) of a metal with that of the other (cathodic half-cell) against which the metal is expected to corrode, it is possible to predict if the metal in question will corrode or not. The simple relation as given by Equation 2.11 is applied for this purpose. In practice, there are other factors, such as applied potential and pH that can also influence corrosion of metal. It will be very useful to obtain information on the corrosion tendency or the stability of a metal under these conditions. Potential-pH diagrams, also called Pourbaix diagrams, are somewhat similar to phase diagrams, outline regions of active corrosion, oxides/hydroxides formation, and thermodynamic immunity for a metal. A simplified Pourbaix diagram for Fe-water system is shown in Figure 2.3 (Pourbaix 1966).

It has to be pointed out that oxides/hydroxides formed on the metal surface can prevent corrosion of the metal, though the mere presence of surface oxide does not guarantee them to be protective. If the oxides/hydroxides protect the underlying metal, then this region is called passive region. This aspect of passivity is very important for engineering alloys to exhibit acceptable resistance to corrosion. As opposed to this region, the thermodynamically immune region ensures complete prevention of metal from corrosion. Therefore, these two regions are identified as anodic protection[7] and cathodic protection regions of metals, respectively. The other area where Pourbaix

---

[7] Anodic protection is employed only in a few systems such as stainless steel/sulfuric acid.

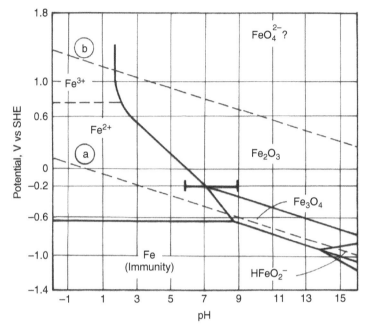

Figure 2.3 Potential-pH diagram (Pourbaix) for iron in water at 25 °C (77 °F). A decrease in pH from 9 to 6 at potential of −0.2 V, which shifts iron from a region of stability to one of active corrosion, is indicated by the solid bar (Pourbaix 1966) (Courtesy CEBELCOR).

diagrams can be useful is in predicting localized corrosion such as pitting and stress corrosion cracking. Figure 2.4 illustrates these aspects with respect to steel in several environments (Jones 1987). The book on "Atlas of electro-chemical equilibria in aqueous solution" is an excellent source for obtaining potential-pH diagrams of metals as well as non-metals (Pourbaix 1966)

2.2.3 Summary

What we have discussed thus far is that

- Corrosion of metallic system is spontaneous and is governed by Gibbs free energy criteria. That is, all corrosion processes must liberate energy.
- While the application of free-energy concept to predict corrosion seems simple, measuring this parameter is difficult.
- Because the corrosion of metallic systems in aqueous environment is electrochemical in nature, it is possible to correlate free energy change to potential that is measurable, albeit with respect to SHE.

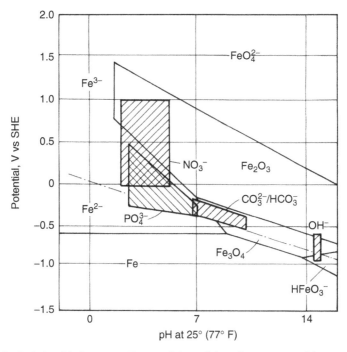

Figure 2.4 Relationship between pH-potential conditions for severe cracking susceptibility of carbon steel in various environments and the stability regions for solid and dissolved species on the Pourbaix diagram. Note that severe susceptibility is encountered where a protective film (phosphate, carbonate, magnetite, and so on) is thermodynamically stable, but if ruptured, a soluble species (Fe^{2+}, $HFeO^{-2}$) is metastable (Jones 1987). (Reprinted with permission of ASM International. All rights reserved www.asminternational.org).

- Once electrode potentials are either measured or determined, the corrosion tendency of a metallic system can be predicted by using the criteria $E_{cell} = E_c - E_a$ and for a spontaneous process E_{cell} has to be positive.
- Pourbaix diagrams are useful in mapping corrosion tendency of a metal against potential and pH.

2.3   KINETICS OF CORROSION

The rate of corrosion (uniform) is an important criterion to be considered for material selection for any structural application. The discussion carried out so far are related to Gibbs free energy change and electrochemical potential that can, at best,[8] predict whether a given metal will suffer corrosion in a given

---

[8]The word "at best" is used to indicate the complexity of environment that is pretty much divorced from ideal conditions employed to illustrate the applicability of above factor for prediction.

environment. This in no way can be useful to predict the life of alloy. For example, look at the EMF series shown in the Table 2.1. It should be obvious from the Table that a structure made of zinc must have higher tendency to corrode than that made of iron. On the contrary, the reverse is true in several environments. This is mainly because of the fact that kinetics (rate of reaction) is governed by other electrochemical factors. Hence, it is imperative to understand kinetics of electrochemical corrosion in order to appreciate how various environmental factors can strongly influence the corrosion rate (CR) of metals. Though the subject is important, a detailed treatment is beyond the scope of this book. For further reading, readers can refer books devoted to electrochemical kinetics of corrosion (Bockris and Reddy 1977; Revie and Uhlig 2008; Fontana 1986; Stansbury and Buchanen 2000).

2.3.1 Description of a Corrosion System

We have seen before that corrosion of a metal involves oxidation from its metallic state. The metallic ions leave the metal lattice, but electrons remain there. The metal surface now is negatively charged. Unless the electrons are consumed by another species, the negative charge on the metal will stop further metallic oxidation. The chemical environment provides species to consume these electrons and so the corrosion proceeds. In such an arrangement, the rate of metal dissolution must be equal to the rate of consumption of electrons (by a chemical species such as H^+, O_2, noble metal ions) to maintain a steady state corrosion. This indirectly implies that the corrosion of a metal depends as much on the kinetics of reduction reaction(s) as on its own oxidation kinetics. In fact, this is the reason why iron corrodes at a faster rate than zinc despite the latter exhibiting lower thermodynamic tendency for oxidation than zinc. A close examination shows that hydrogen reduction kinetics on zinc surface is much lower than when it occurs on iron surface despite the fact that the driving force for hydrogen evolution on zinc is more than that on iron (Look at the E_{cell} value). The point to be emphasized here is that understanding cathodic and anodic reaction kinetics is very important. Therefore, learning electrochemical kinetics is a prerequisite (i) to develop alloys that can exhibit high corrosion resistance,[9] (ii) to examine the corrosivity of any environment, and (iii) to even carry out corrosion testing and monitoring.[10]

It will be very instructive to picturize how the potential and current at a metal/environment interface vary from the moment a metal is immersed in an environment to the time metal/environment interface reaches a steady state. Figure 2.5 illustrates this.

---

[9]See the Chapter 4 on "Materials of Construction for chemical process industries" for more details.
[10]See Section 2.4 "Corrosion Evaluation and Monitoring" for more details.

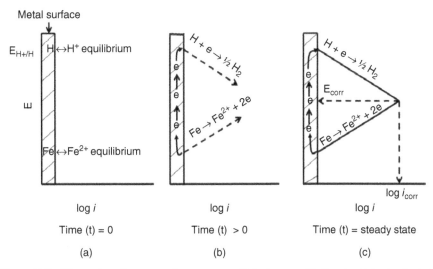

Figure 2.5 Illustration of events and parameters which change with time to attain steady state corrosion. See the text for details regarding (a), (b) and (c).

At the first instance, the metal surface establishes two independent equilibria, one with metal and its metal ions (say Fe^{2+}/Fe) and the other with chemical species of the environment (say H^+/H) (see Fig. 2.5a). Because the two equilibrium potentials lie apart from each other and the metal is a conductor, with least resistant to flow of current, the metal tries to achieve a uniform potential, through the flow of electrons. In this process, the electrochemical equilibrium with higher potential (H^+/H) drifts towards lower potential, by accepting electron (reduction), while the other equilibrium with lower potential (Fe^{2+}/Fe) drifts towards higher potential by releasing electrons. This increases the rate of oxidation and reduction reactions (Fig. 2.5b). A steady state between the two reactions is reached based on the two following criteria (i) the metal must attain a uniform potential, called, corrosion potential, E_{corr}, and (ii) the rate of oxidation must equal the rate of reduction (Fig. 2.5c), the current flowing through the metal at this stage is called corrosion current (I_{corr}). When normalized with respect to area, it is called corrosion current density (i_{corr}). Figure 2.5c is called Evans diagram.

2.3.2 Predicting Corrosion

Thermodynamics is a state property, while the kinetics relies on the route through which the transition occurs. Various steps involved during corrosion of a metal were briefly described in the previous section. If we can quantify the parameters that describe the route, then we can predict the corrosion through a quantitative relation.

On examination of Figure 2.6, the Evans diagram, the following characteristics of cathodic and anodic reactions become obvious.

1. The potential and current are related in a simple manner through over voltage η ($=E_{app} - E_{eqm}$).[11] Over voltage η and log i curves bear a first order linear relation as given below.

$$\eta_a = -a + \beta_a \log i \qquad (2.12)$$

$$\eta_c = a - \beta_c \log i \qquad (2.13)$$

Where, "a" is an intercept,[12] which is equal to $\frac{RT}{nF} \log i_o$ and β_a and β_c are anodic and cathodic Tafel slopes, respectively. Such a relation was first established by Tafel through experiments, though Butler-Volmer later gave a theoretical basis for this equation. Figure 2.6 shows how each of the Tafel slopes can be obtained from the diagram. Given the Tafel slopes, equilibrium potentials, and exchange current density for cathodic and anodic reactions, it is possible to compute the corrosion current density. It is also possible to determine the same through experimental means.

2. It is now clear both from the diagram and the equation that larger the difference between E_{eqm} (on either side of E_{corr}) and E_{corr}, also called as

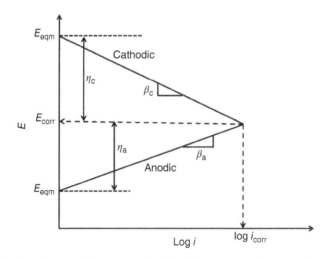

Figure 2.6   Evans diagram illustrates various kinetic parameters that control corrosion rates.

[11] Deviation from the equilibrium state is called polarization.
[12] We haven't discussed the role of i_o, as it calls for a detailed treatment of electrochemical interface. For details refer the book by Bockris and Reddy (1977).

overvoltage, higher will be the i_{corr}. Furthermore, the flow of corrosion current is primarily due to the two equilibria deviating to become either anodic or cathodic reactions.

3. While increasing η_c and/or η_a increases i_{corr}, a similar increase in β_c and/or β_a decreases i_{corr}.
4. While it is not possible to change E_{eqm} and i_o, as they depend on the metal and environment, it is quite possible to alter β values and thereby control corrosion. Modification of β_a and/or β_c through inhibitor addition is indeed a practical measure of controlling corrosion.
5. It should be pointed out that E_{corr} of a metal/alloy gives a better picture of relative nobility of metals than standard potential and also forms a basis for predicting galvanic corrosion or dissimilar metal corrosion. This aspect will be elaborated later under the topic galvanic corrosion.

2.3.3   Passivity

The potential-pH diagram brings out the regions where a metal has an ability to form oxide/hydroxide on its surface. However, the extent to which these oxides can protect a metal from corrosion cannot be obtained from such diagrams. Electrochemical kinetic diagrams are useful in this context.

As per the Tafel equation (Equation 2.12), a steady increase in anodic polarization (η_a) should result in enhanced anodic current and hence metallic dissolution. However, some systems (metal/environment) exhibit initial increase in anodic dissolution, and later a reduction in the dissolution rate when η_a is increased. The dissolution rate then remains almost constant despite an increase in η_a—called a passive state—and after reaching high anodic (positive) potentials, the metal again exhibits high dissolution rate. Such transitions are sketched in Figure 2.7.

The characteristics of anodic polarization curve can be described by the parameters such as critical current density for passivation (i_c), passivation potential, (E_p), passive current density (i_p), and pitting (E_{pit}) or transpassive (E_{tp}) potential. These parameters are important as they reflect not only the CR but also the tendency of the metal to resist localized corrosion such as galvanic corrosion, pitting corrosion, crevice corrosion, erosion corrosion, and stress corrosion cracking in a given environment. Table 2.2 summarizes how these parameters can affect the corrosion resistance of a metal/alloy. The above description indicates how a passive metal having a characteristic anodic behavior is able to resist various forms of corrosion. Nevertheless, the role of cathodic kinetics needs no emphasis, as it can either promote the corrosion resistance or spoil its virtues. Just consider three types of cathodic reactions as marked in Figure 2.8a to illustrate the above argument. (i) The type I cathodic reaction does not enable the metal to achieve its passivity, as

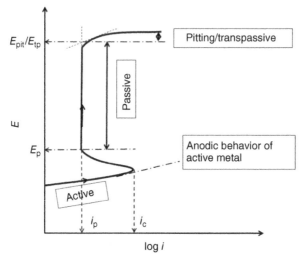

Figure 2.7   Schematic of anodic polarization of a metal exhibiting active-passive-transpassive (pitting) regions. See the text for the parameters shown in the diagram.

TABLE 2.2   Relation Between Electrochemical Parameters and Corrosion

Parameter	Effect on Corrosion
i_p, decrease	High resistance to uniform, galvanic, erosion, stress corrosion cracking
i_c, decrease	High resistance to crevice corrosion, stress corrosion cracking, high passivity
E_p, low	High passivity
E_{pit}, high	High resistance to pitting

E_{corr} lies in active region (Fig. 2.8b). (ii) The type II cathodic reaction plays the spoiled sports and enhances the CR, as E_{corr} lies in the active–passive transition zone (Fig. 2.8c). (iii) Type III cathodic reaction helps to keep the metal in the passive state and so lowers the i_{corr}, lower than that corresponds not only with respect to the type II but also type I cathodic reaction (Fig. 2.8d). It is in this respect low i_c and low E_p of a metal are very important. In such cases even type I and type II cathodic reactions can enable an alloy to attain passivity, spontaneously.

2.3.4   Summary

Electrochemical kinetics forms the basis for predicting/measuring CRs of metals. A typical corrosion system will have anodic (oxidation of metal) and

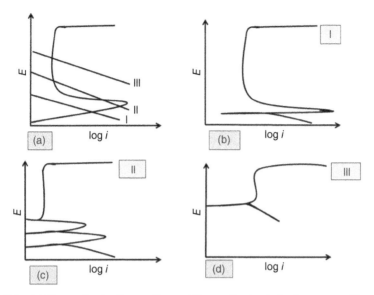

Figure 2.8 (a) Types of cathodic reaction and the resultant polarization curves. Figures (b), (c), and (d) are polarization curves, respectively, for Types I, II, and III cathodic reactions.

cathodic reactions which are independently governed by their own kinetic parameters such as overvoltage (η) and Tafel slopes (β) and exchange current density (i_o). In addition, the rates of anodic and cathodic reactions are the same, under steady state conditions, irrespective of metal and environment. Based on the steady state criteria, it is possible to compute E_{corr} and i_{corr} values, if the above kinetic parameters are provided. On a more practical side, it is possible to measure both E_{corr} and i_{corr} both in the laboratory and on the field.

Active metals, under some specific conditions, corrode at a low rate by forming passive film. The passivation characteristics of such metals can be obtained based on their anodic polarization curves. The parameters such as passive current density, (i_p), critical current density (i_c), passive potential (E_P), and pitting potential (E_{pit}) can be used to compare corrosion, crevice corrosion, and pitting corrosion tendency of various metals.

2.4 CORROSION EVALUATION AND MONITORING

Most techniques of laboratory corrosion evaluation and onsite corrosion monitoring may operate based on similar principles but have differing objectives. The former is used for evaluation of corrosion resistance of materials, corrosive nature of environment, and efficiency of protection techniques such as inhibitors and coatings and even corrosion mechanisms. In this case, the test conditions are reasonably simple in nature. On the other hand, onsite corrosion monitoring techniques are complex by nature. Because of complexities, only robust techniques can be applied even if they may not be as accurate as those used in the laboratory in determining CRs. Detailed information on laboratory corrosion testing can be obtained from well known books such as those by Kelly et al. (2002) and the book series Corrosion Testing Made Easy with the series editor Syrett, B. C and published by NACE American Society of Testing of Materials standards provide details of the experimental procedures for various types of corrosion and NACE report (T-3T-3) give an overview of corrosion monitoring techniques. For the convenience of the readers, some of the important standards are listed at the end of this chapter.

Only those techniques that are widely employed and provide reasonable reliability in data interpretation (especially for online monitoring) are described. The techniques can be broadly classified as (i) electrochemical and (ii) non-electrochemical techniques and discussion is made as per this classification.

2.4.1   Electrochemical Techniques

Using these techniques corrosion current and/or corrosion potential are measured as briefly outlined below

- Corrosion current measurement.
- Linear polarization resistance (LPR).

In this technique two electrodes consisting mostly of similar material are used. The polarization resistance (R_p) of a corroding system can be determined, by measuring the current, provided the applied potential (ΔE) between two metal electrodes is small.

$$R_p = \frac{\Delta E}{\Delta I} = \frac{\beta_a \beta_c}{2.303 i_{corr}(\beta_a \beta_c)} \tag{2.14}$$

β_a and β_c are Tafel slopes as described earlier in Figure 2.6. In practice, β_a and β_c are not measured, but are assumed to lie around 100 mV/decade. As the applied potential on the electrode is small, it does not suffer severe corrosion during the test. Hence, this test is considered as non-destructive in nature. Though this technique can be used in a laboratory, because of its simplicity and versatility, one needs to find wider application for online corrosion monitoring of units such as pipelines and cooling water systems.

2.4.1.1   *Potentio/Galvano Dynamic Technique*   It is possible to generate complete Evans diagram for anodic and cathodic polarization curves of a metal electrode, exposed to corrosive environment, such as those shown

earlier in Figure 2.6. This technique is capable of determining corrosion potential, corrosion current density, Tafel slopes, passivation, and pitting parameters. This is a very important technique used in the laboratory for evaluation of materials as well as for research for understanding corrosion mechanisms, but is not versatile enough for application in the field. The instruments for this technique namely potentio/galvano stats are more expensive than those used for LPR technique.

2.4.1.2   *Zero Resistance Ammeters* The tendency for galvanic corrosion between dissimilar metals depends on how far apart they are with respect to their galvanic potentials (See Section No 3.3 Chapter III). However, the galvanic CR suffered by an active metal depends on the galvanic current that flows between the two dissimilar metals. This can be directly measured using a zero-resistance ammeter (ZRA). It is possible to convert a potentiostat into ZRA. ZRA can serve as online real time technique to determine how process variables can affect CRs of dissimilar metals employed in the process.

Relation Between CR and Corrosion Current Density Corrosion current can be related to CR through Faraday's law. Once the corrosion current is measured, the CR of the metal can be determined. The following simplified equation correlates the two.

$$\text{Corrosion rate} = K\frac{a \times i}{n \times \rho} \tag{2.15}$$

where, a, i, n, ρ, and K, respectively, correspond to atomic weight in g, current density in $\mu A/cm^2$, oxidation state of metal, density (g/cm^3), and constant which equals 0.129 and 0.00327, respectively, for CRs in mpy and mm/y.

Potential Measurements The corrosion potential (E_{corr}), also known as open circuit potential and rest potential, of a corroding metallic system as opposed to corrosion current is an indirect measure of the tendency of the corrosion it experiences. The technique is simple, as just by using the popular reference electrodes, such as saturated calomel, silver/silver chloride, or copper/saturated copper sulfate electrodes, E_{corr} can be measured. This technique can be employed in a laboratory as well as on the field for online monitoring. Notably corrosion potential is also an indication of the level of O_2, biocides, and inhibitors present in the system that affect the corrosion tendency of a metal. Potential monitoring is an important corrosion monitoring technique for structures protected by cathodic protection.

2.4.2 Non-electrochemical Techniques

Weight loss, electrical resistance (ER), ultrasonic techniques as well as estimation of corrosion products give a measure of CR of metals.

2.4.2.1 Weight Loss This is by far the simplest technique used for CR determination either in a laboratory or in a plant, as it involves no assumption in terms of relation between the measured parameter and the CR. The relation between the CR and the metal loss is given by the following equation:

$$\text{CR} = k \frac{W}{dAt} \tag{2.16}$$

W, d, A, t, and k correspond to weight loss in g, specimen density in g/cm^3, specimen area in cm^2, time of exposure in hour, and constant, respectively. The value of k can be either 5.34×10^5 or 8.76×10^4 depending on whether the CR unit is mpy or mm/y.

The technique involves determining weight loss of a metallic coupon, of known area exposed to a definite period of time. The duration of exposure is inversely proportional to CR (≤ 2000/mpy) and so the technique is time-consuming, especially when the metals corrode at low rate. Hence, this technique cannot be used as online real time measurement. It is also important that the coupons are properly cleaned in order that the oxide scales formed on the coupons do not interfere with the actual metal loss. The Standard ASTM G1-03 describes cleaning procedures for various metallic systems.

2.4.2.2 Electrical Resistance The electrical resistance (R) of a conductor is given by

$$R = \rho \frac{L}{A} \tag{2.17}$$

where ρ, L, and A correspond to resistivity, length, and area of a conductor.

Here metal probes of wire, tube, or strip are used. On corrosion, diameter or the thickness of the probe, depending upon which of the three being used, decreases. This as per the above equation reflects in R. So the corrosion monitoring involves measuring the ER of the exposed metallic probe (wire, tube, and strip) with time.

In order to compensate the influence of temperature on resistivity, the change in ER of the exposed metal can be compared with an identical probe (reference) that is not exposed to the environment. The technique can be used for online as well as off-line real time measurements. Another advantage of this technique over LPR, potentio/galvano static, and E_{corr} measurement techniques is that it does not necessitate complete immersion of probes in the environment as well as the environment need not be conducting. Hence, this is an ideal technique for measuring dew point corrosion on location such as over-heads of crude distillation columns.

2.4.2.3 Ultrasonic The thickness of a solid object can be measured using this technique. The time taken for a sound wave of known velocity to travel across the solid is a measure of the thickness of the solid. The technique is very versatile for inspection of plants as well as online corrosion monitoring. This technique is non-intrusive in nature, though direct access of surface of the component is needed.

2.4.2.4 Reactants and Products Analysis Corrosion reaction involves reduction of species such as H^+ to H_2 and oxidation of metal to its ions. So, by measuring H_2 and determining metal ion concentration in the stream, corrosion experienced by a component can be determined. The former is done using hydrogen probes and the latter using some spectrometers (atomic absorption, inductively coupled plasma, etc.) or ion selective electrodes. For corrosion monitoring of oil and gas pipelines, hydrogen probes have been found to be more suitable. The metal ion analysis technique, on the other hand, can be used for assessing corrosion in cooling water and boilers to get an idea on the health of the system.

It is known that the corrosion tendency of a metal depends on environment pH and dissolved oxygen (DO). So, pH and DO meters find applications in online monitoring of corrosivity of environments. For example, both these meters are used to monitor boiler corrosion while the pH meter is used in the case of cooling water systems.

REFERENCES

Bockris, J. O'. M. and Reddy, A. K. N., (1977), *Modern Electrochemistry*, Vol. 2, Plenum, New York.

Fontana, M. G., (1986), *Corrosion Engineering*, McGraw Hill, New York.

Jones, R. H., (1987), *Stress Corrosion Cracking*, Vol. 13, 9th Edition, ASM Handbook, p. 152.

Kelly, R. G., Scully, J. R., Shoesmith, D. W. and Buchheit, R. G., (2002), *Electrochemical Techniques in Corrosion Science and Engineering*, Marcel Dekker, Inc., New York.

NACE Technical Committee report Task Group T-3T-3. Techniques for Monitoring Corrosion and Related Parameters in Field Applications.

Pourbaix, M., (1966), *Atlas of Electrochemical Equilibria in Aqueous Solutions*, Pergamon Press, Oxford.

Revie, W. R. and Uhlig, H. H., (2008), *Corrosion and Corrosion Control: An Introduction to Corrosion Science and Engineering*, 4th Edition, John Wiley & Sons, Inc.

Stansbury, E. E. and Buchanan, R. E., (2000), *Fundamentals of Electrochemical Corrosion*, ASM International, Ohio.

Syrett, B. C., Corrosion Testing Made Easy, NACE International Publication.

SOME USEFUL STANDARDS

NACE Standard TM0169-2000. Laboratory corrosion testing of metals.

NACE Standard RP0189-2002. On-line monitoring of cooling waters.

NACE Standard RP0502-2002. Pipeline external corrosion direct assessment methodology.

ASTM G3-89. Standard practice for conventions applicable to electrochemical measurements in corrosion testing; 2004.

ASTM G5-94. Standard reference test method for making potentiostatic and potentiodynamic anodic polarization measurements; 2004.

ASTM G96-90. Standard guide for on-line monitoring of corrosion in plant equipment (electrical and electrochemical methods); 2001.

G102-89. Standard practice for calculation of corrosion rates and related information from electrochemical measurements; 2004.

3

FORMS OF CORROSION

3.1 INTRODUCTION

One of the main tasks of corrosion failure analysis is to decode the failure mechanism(s) of the component in question from available data. The electrochemical corrosion processes discussed in the previous chapter are meant to understand:

(a) How to predict the stability of metals against corrosion and
(b) How the cathodic and anodic reactions in any corrosion process will behave.

But, what makes actual corrosion failures more complex is the way these two reactions are (a) distributed over the metallic surface and (b) affected by external variables such as bimetallic joints, applied stresses, flow condition of the environment, to mention a few. Such variations lead to different mechanisms of corrosion or simply called different forms of corrosion. Each corrosion failure leaves behind some sort of fingerprints that characterize a corrosion failure. It is necessary to understand these aspects before starting to investigate corrosion failures. In addition to this, it is imperative to understand what factors govern these failures and how they can be controlled for providing appropriate recommendations. A cursory look at the chapter on case

Corrosion Failures: Theory, Case Studies, and Solutions, First Edition. K. Elayaperumal and V. S. Raja.
© 2015 John Wiley & Sons, Inc. Published 2015 by John Wiley & Sons, Inc.

studies (Chapter 6) will indicate that an industrial component can suffer any of the following corrosion failures, though the frequency of an occurrence of any of these failures may vary with the type of industries.

(a) Uniform corrosion
(b) Galvanic corrosion
(c) Pitting corrosion
(d) Crevice corrosion
(e) Intergranular corrosion
(f) Selective dissolution/selective attack
(g) Flow-assisted/erosion/cavitation corrosion
(h) Stress corrosion cracking
(i) Hydrogen-assisted failure
(j) High temperature corrosion

All these forms of corrosion failures are briefly discussed below. For a detailed understanding, readers can refer an earlier publication by one of the authors (Raja 2008).

3.2   UNIFORM CORROSION

In a corrosion process, the locations of anodic and cathodic reactions are separated at microscopic levels on a metallic surface. However, a metal tends to corrode uniformly when the anodic and cathodic reactions exchange their positions with time in equal measures. If their positions are fixed over a time period either at microscopic or macroscopic level, the metallic surface suffers localized corrosion.

The rate of uniform corrosion and hence the life of any component is predictable. The corrosion rate is measured by units such as mpy (mils per year) and mm/y (millimeter per year). Some examples of such corrosion are atmospheric corrosion of steel structures, automobile body corrosion, and steel tank corrosion due to sulfuric acid. It is worthwhile to point out the fact that even steel surfaces suffered by the so-called uniform corrosion can more often appear uneven tempting one to conclude it to be due to pitting corrosion. Pitting corrosion, on the other hand, is characterized by deep attacks as will be discussed later. One or more of the following methods can be used to control uniform corrosion

(a) Selection of better materials
(b) Protective coatings

(c) Inhibitors

(d) Cathodic protection

3.3 GALVANIC CORROSION

On several occasions, dissimilar metals are used to fabricate a component because of functional requirements. Heat exchangers (shell and tubes with different alloys) and pipelines (pipe and valve with different alloys) are two such examples. When two metals are electrically in contact with each other and exposed to the environment, they develop two different potential, called galvanic potential, with respect to the environment; therefore an electrical current starts flowing between the two metals. The metal having higher potential becomes cathodic, while the other is rendered anodic. In this case, the latter becomes a permanent anode causing localized corrosion (recall the definition of localized corrosion described in the above section). This type of corrosion is called galvanic/dissimilar metal corrosion. Galvanic potentials of various metals given in Table 3.1 give an idea of relative nobility of various metals and alloys in seawater.

The above point is illustrated below: Carbon steel when comes into contact with alloys like stainless steel or brass it corrodes much faster than its normal rate in aqueous solutions. Carbon steel is active to stainless steel/brass and the latter is relatively noble to the former. On the other hand, when zinc comes into contact with steel the latter becomes relatively nobler than the former. In fact, this is the principle behind cathodic protection of steel with zinc anodes, zinc coatings, and zinc-rich paints. A typical galvanic corrosion attack of steel in contact with stainless steel is illustrated in Figure 3.1.

3.3.1 Factors Affecting Galvanic Corrosion

- Galvanic Potentials: Any metal/alloy in the series (Table 3.1) when comes into contact with a metal that lies above in the galvanic series becomes active, but when it comes into contact with the one below in the series, it comes noble. Larger the difference in galvanic potential, higher is the tendency for galvanic corrosion of the active metal.

- Chemical nature of the environment: Higher the conductivity of the environment, larger is the galvanic corrosion.

- Electrochemical polarization characteristics: If the active metal becomes passive, either through passivity or by a protective corrosion product film formation, galvanic corrosion would not proceed. This is why sometimes the zinc anodes applied for cathodic protection of steel are not

TABLE 3.1   Galvanic Series of Various Metals Exposed to Seawater

Active End	Magnesium
(−)	Magnesium alloys
↑	Zinc
|	Galvanized steel
|	Aluminum
|	Aluminum 6053
|	Alclad
|	Cadmium
|	Aluminum 2024 (4.5 Cu, 1.5 Mg, 0.6 Mn)
|	Mild steel
|	Wrought iron
|	Cast Iron
|	13% Chromium stainless steel
|	Type 410 (active)
|	18-8 Stainless steel
|	Type 304 (active)
|	18-12-3 Stainless steel (active)
|	Type 316 (active)
|	Lead-tin solders
|	Lead
|	Tin
|	Muntz metal
|	Manganese bronze
|	Naval brass
|	Nickel (active)
|	76 Ni-16 Cr-7 Fe alloy (active)
|	60 Ni-30 Mo-6 Fe-1 Mn
|	Yellow brass
|	Admirality brass
|	Aluminum brass
|	Red brass
|	Copper
|	Silicon bronze
|	70:30 Cupro nickel
|	G-Bronze
|	M-Bronze include
|	Silver solder
|	Nickel (passive)
|	76 Ni-16 Cr-7 Fe
|	Alloy (passive)
|	67 Ni-33 Cu alloy (monel)
|	13% Chromium stainless steel

(continued)

TABLE 3.1   (*Continued*)

Active End	Magnesium
|	Type 410 (passive)
|	18-8 stainless steel
|	Type 304 (passive)
|	18-12-3 stainless steel (passive)
↓	Type 316 (passive)
(+)	Silver
NOBLE or	Graphite
PASSIVE END	Gold
	Platinum

(Reprinted with permission from *ASTM G82-98* (*2009*) *Standard Guide for Development and Use of a Galvanic Series for Predicting Galvanic Performance*, copyright ASTM International, 100 Barr Harbor Drive, West Conshohocken, PA 19428. A copy of complete standard may be obtained from ASTM International, www.astm.org)

Figure 3.1   Galvanic corrosion attack of carbon steel bolt in contact with 12% Cr steel (AISI 410) stainless steel plate in hydrocarbons contaminated with traces of salts. This is ray lamp of a distillation column. (Source: During (1988). Reproduced with permission of Elsevier.).

effective in certain environments in which they get passivated. It is also possible that some cathodes are less effective in promoting galvanic corrosion on its anodic counterpart. Ti-alloys are poor cathodes, though they are very noble to many metals/alloys.

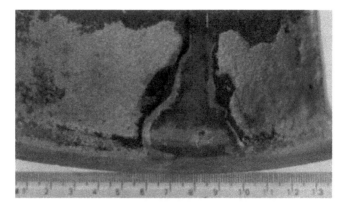

Figure 3.2   Severe microgalvanic attack on smaller areas of the weld seam (type 304L SS) by large areas of the parent metal (super austenitic stainless steel) in an organic solvent containing sulfuric acid as a carrier medium. (Courtesy: Elayaperumal, in "Corrosion Failures in Process Industries in India: Statistical Analysis and Case Studies" in Special Supplement 2009, Corrosion Reviews Special Issue: India, England: Freund Publications, 2009, p 57, case study 8.)

- Area Effect: If the exposed surface area of the cathodic metal is much more than that of the anodic metal, the extent of acceleration of corrosion of the anodic/active metal is much higher than that corresponding to the coupling of equal areas of both the metals. A practical example from chemical process industry is the observed accelerated corrosion of small area carbon steel tube sheet in contact with large area stainless steel tubes in shell and tube heat exchangers/condensers. Similarly severe galvanic attack on smaller areas of weld seams (if filler wires happen to be active to the parent metals) by large areas of the parent metals (Figure 3.2).

- Distance Effect: If the conductivity is very high, the galvanic attack on active metal extends to a large distance from the point of electrical contact, otherwise the affected portion is only very close to the jointed area. This is commonly seen on the interior of butt-welded pipelines if the weld metal composition is more noble/cathodic as compared to the base metal composition.

3.3.2   Controlling Galvanic Corrosion

1. Choose metals closer in galvanic series
2. Use a third metal that is active to the active member of the galvanic couple
3. Coat both the anodic and cathodic members of the galvanic couple. If not, coat the cathodic member of the couple and never just the anodic member
4. Add inhibitors to lower the aggressiveness of the environment.

3.4 PITTING CORROSION

Pitting corrosion is one of the most serious forms of corrosion that can severely damage engineering materials with undesirable consequences. Generally, pitting corrosion occurs on alloys when they are in the passive state. Some of the alloys/metals such as stainless steels and Al, Ti, Zr, and their alloys are known to passivate in a wide-ranging industrial environments. Iron and steel, though do not normally passivate, can exhibit passivity in some environments and therefore can suffer pitting under these conditions. Pitting can be more severe than uniform corrosion. For example, a pipeline made of Type 304 SS can leak earlier than a carbon steel or cast iron pipeline when used for carrying seawater. Hence, this topic assumes a great significance.

3.4.1 Pitting Process and Pitting Morphology

Pitting corrosion of an alloy is caused by specific anions such as chloride damaging the passive films at microscopic levels. As an illustration hemispherical pits formed on type 316 stainless steel (ss) and irregular pits formed in an aluminum-graphite composite are shown in Figure 3.3.

Pitting process follows two steps namely initiation and propagation. Among the various mechanisms proposed for initiation, the most popular and accepted one is that of the absorption of aggressive anion, (i) either in preference to oxygen on an active metal surface or (ii) on a passive metal surface, causing localized "pitting" corrosion.

A passive metal dissolves at a rate given by its passive current density which is of the order 10^{-6} A/cm^2, while in the pit the dissolution current density can rise up to 10 A/cm^2. Thus, the dissolution rate of a pit could be a million times higher than that of passive metals. Similarly pits encounter an entirely different environment than what is encountered by passive surface. Growing pits are associated with low pH, high chloride (or halides) concentration, and precipitation of salts, which prevent repassivation of pit surface.

Figure 3.3 shows two types of pits, hemispherical and irregular morphologies. However, in practice pitted metallic surface can exhibit several different morphological pits. ASTM Practice G-46 describes different pit morphologies a metal/alloy can exhibit (Figure 3.4). Chemistry, microstructure, and the nature of inclusions of the alloy influence the pit morphology.

3.4.2 Factors Affecting Pitting Corrosion

Pitting tendency of a metal or alloy is governed by many factors. They are

1. Electrochemical
2. Metallurgy

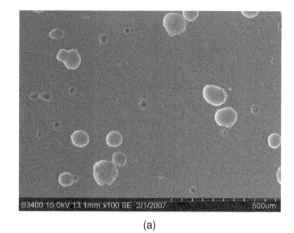

(a)

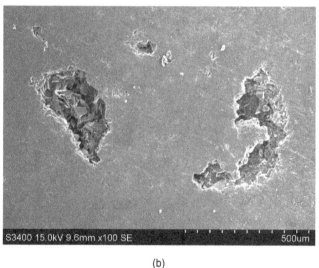

(b)

Figure 3.3 (a) Hemispherical pits formed on type 316 ss after anodic polarization in 3.5 wt% NaCl solution and (b) Irregular pits formed in Al-SiC composite after anodic polarization in the same medium.

3. Environment chemistry
4. Surface condition

The electrochemical parameters that reveal pitting corrosion tendency of a metal were described earlier in Section 2.3.3. High pitting (E_{pit}) and repassiva-tion (E_{prot}) potentials are indicative of an alloy resistant to pitting. In fact these parameters form the basis of material selection to resist pitting corrosion. It will not be out of place to mention the fact that a more simplistic parameter namely "critical pitting temperature" is also used as a parameter in selecting

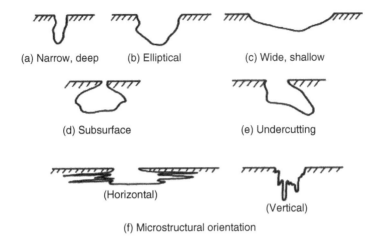

(a) Narrow, deep   (b) Elliptical   (c) Wide, shallow

(d) Subsurface   (e) Undercutting

(Horizontal)

(Vertical)

(f) Microstructural orientation

Figure 3.4   Possible variations in the pit morphology, seen through sample cross-section, are shown. (Reprinted with permission from ASTM G46. Copyright ASTM International, 100 Barr Harbor Drive, West Conshohocken, PA 19428.)

pitting-resistant alloys. Critical pitting temperature is defined as the lowest temperature below which pitting does not occur on a metal when tested as per ASTM standard G48.

Regarding the metallurgical aspects, it is well known that chemical composition, microstructure, and micro and macro segregation exert a profound influence on pitting tendency or otherwise of an alloy. For example, Cr, Ni, Mo, W, and N play a dominant role in promoting pitting resistance. On the other hand, weld-fusion zone is more prone to pitting than the unwelded wrought alloys (due to microsegregation of the alloying elements of stainless steel especially Cr and Mo in the weld-fusion zone). Manganese sulfide (MnS) inclusions are unstable especially in acidic conditions and found to be a source for pitting in stainless steels. Sensitization causes a steep drop in chromium content of the grain boundary region (refer the section on intergranular corrosion for details). This becomes responsible for poor pitting resistance of stainless steels when subjected to improper heat treatment and/or improper welding. A detailed treatment of various factors affecting pitting corrosion can be found in Chapter 4 on Materials Selection.

Though pitting is a severe form of corrosive attack, the favorable aspect of pitting is that only certain types of anions are responsible for pitting. These ions can be further assisted by other ions—both anions and cations. Chlorides, commonly found in normal water sources, pit stainless steels and aluminum alloys. The severity of pitting increases with the concentration of the chloride ions and temperature and decreases with a rise in pH. Chlorides bring down the pitting potential. The species such as dissolved O_2, noble metal cations

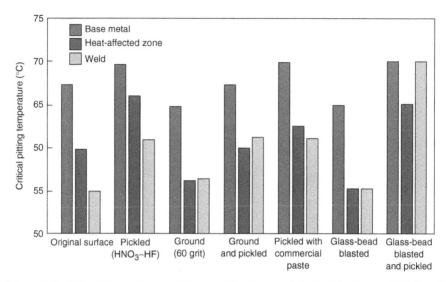

Figure 3.5 Role of the surface condition on pitting is exemplified in this diagram. Pickling of stainless steel increases the critical pitting temperature in FeCl3 in the base metal, heat-affected zone, and welds areas. Mechanical cleaning treatments that are performed without a subsequent pickling treatment decrease the critical pitting temperature. (Reprinted with permission of ASM International. All rights reserved www.asminternational.org.)

$(Cu^{2+}, Hg_2^{2+}, Ag^+)$, and ions having high equilibrium potentials (Fe^{3+}) can raise E_{corr} of the passive metal above E_{pit} and thereby promote severe pitting in presence of chloride ions. Ions such as SO_4^{2-}, OH^-, and ClO_3^-, CrO_4^{2-}, and NO_3^- are found to reduce the pitting tendency of stainless steels.

Surface condition influences the passive film stability and so affects pitting tendency of any metal. Figure 3.5 clearly demonstrates that the high pitting resistance of stainless steels can be achieved by

(a) Lowering the surface roughness
(b) Removing the mill scale and
(c) Passivating the stainless steels (Tuthill and Avery 1992).

ASTM designation A380-78 is a recommended practice for passivation of stainless steels.

3.4.3 Controlling Pitting Corrosion

1. Choose alloys resistant to pitting, such as stainless steels high in Mo and N

2. Control the environment wherever possible by lowering/eliminating the ions, such as chlorides, responsible for pitting.

3.5 DIFFERENTIAL AERATION-ASSISTED CORROSION (CREVICE, UNDER DEPOSIT, AND WATER-LINE CORROSION)

Passive metals exposed to corrosive environment when subjected to differential aeration develop electrochemical cell. The location having easy access to air/oxygen readily passivates, whereas the region depleted of air/oxygen looses the tendency to passivate. The depleted region becomes a permanent anode and the surrounding area a permanent cathode. Such a situation occurs under the following conditions/locations:

1. Flange and lap joints
2. Interface of rolled tube ends
3. Threaded joints
4. Riveted areas
5. O-rings and gaskets
6. Weld fusion zones with inadequate fusion
7. Under deposits
8. Water-line area

3.5.1 Characteristics of Differential Aeration Corrosion

1. Occurs only when a metal passivates (note: this depends on alloy as well as the environment)
2. Crevice deposits must exhibit restricted flow of the electrolyte (If crevice gap is enhanced differential aeration does not occur and crevice corrosion will not be initiated)
3. The growth of crevice causes a drop in pH and a rise in chloride level and so the growth becomes autocatalytic. That means, crevice once stabilized will continue to grow even if the differential aeration ceases to operate
4. Crevice corrosion is more severe than pitting corrosion. A typical crevice attack under a gasket is shown in Figure 3.6. It is clear that the attack occurred at the metal-gasket interface.

Figure 3.6 Photograph of a part of heat-exchanger sheet affected by crevice corrosion. The carbon steel shell was weld-cladded with SAF 2507 alloy. Note, only the gasket area of the shell suffered from attack; shell side was exposed to seawater. (Source: Raja (2008). Reproduced with permission of Narosa Publishing House.)

3.5.2 Factors Affecting Differential Aeration Corrosion

The following factors influence crevice corrosion tendency of a metal

1. Electrochemical
2. Crevice gap
3. Environment
4. Metallurgy
5. Design

Low passive potentials and low critical current density increase the differential aeration corrosion resistance of metals. Similarly, a rise in crevice gap or sealing the crevice can eliminate crevice corrosion. Though differential aeration is the cause for initiating crevice corrosion, the severity of corrosion depends on species such as chloride and H^+ present in the environment. Increase in their level raises the tendency of crevice corrosion attack. In addition to the chemistry and temperature of the environment, stagnancy and fouling can largely contribute to the crevice corrosion. Corrosion under the fouled areas is termed as underdeposit corrosion.

Differential aeration corrosion is known to occur in passive alloys in much the same way pitting corrosion does. Interestingly, the alloys which exhibit high resistance to pitting also exhibit high resistance to differential aeration corrosion. Thus for an alloy to be resistant against crevice corrosion, it can be that the alloy must either exhibit extremely high passivity or no passivity at all. Thus, metals such as Ti, Zr, and Ta and their alloys exhibit high crevice corrosion resistance in seawater due to high passivity, while carbon steels do not suffer much from such an attack, as they do not passivate in seawater.

Fouling and stagnancy of an environment in a vessel/reactor cause under-deposit and waterline corrosion, respectively. Both of these, to a large extent, depend on the reactor design. Any design that could avoid dead ends and provide good drainage points can minimize crevice corrosion. Wherever possible, joints that encourage crevice formation need to be replaced with weld joints. It should also be emphasized that welds should be of high quality so that this by itself is not a source of crevice corrosion.

3.5.3 Differential Aeration Corrosion Control

(a) Avoid flange joints, instead weld the joints
(b) Use gaskets that do not absorb water
(c) Keep the system clean and free from deposition
(d) Obtain sound welds, free from porosity and cracks
(e) Design systems for complete drainage, on occasions of shutdown
(f) Choose materials having high resistance to crevice corrosion

3.6 INTERGRANULAR CORROSION

Grain boundaries in alloys can be regions of localized attack. Chemical segregation in grain boundary, depletion of passivating elements along the grain boundary region, and preferential precipitation of phases along the grain boundary are the most important causes of intergranular corrosion (IGC) in metals and alloys. Among all these, intergranular corrosion of stainless steels has a major impact on fabrication of stainless steel components through welding. Hence, this aspect is discussed more in detail than others in this section.

3.6.1 IGC of Stainless Steels

Stainless steels are known to suffer from intergranular corrosion if they are exposed to certain high temperature range. The critical temperature over which a stainless steel becomes sensitive to intergranular corrosion depends on the type of stainless steel and its composition. For austenitic stainless steel, this range lies between 450 and 850 °C. The reason for the loss in corrosion resistance of stainless steels is normally explained as below.

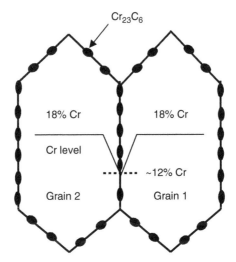

Figure 3.7 Schematic diagram showing $Cr_{23}C_6$ precipitations and chromium depletion at grain boundaries.

3.6.1.1 Sensitization Phenomena

The stainless steels, especially, the austenitic types comprising 3XX series, are produced with C levels beyond its room temperature solubility limit. So under thermodynamically equilibrium conditions, these stainless steels should contain both austenite and chromium carbide ($Cr_{23}C_6$) phases. To prevent the formation of $Cr_{23}C_6$, stainless steels are solution heat treated at 1050 °C and quenched, cooled fast, to a low temperature. This treatment retains C in the soluble form and prevents the formation of $Cr_{23}C_6$. But long exposure to the 450–850 °C temperature range enables chromium carbide formation, which occurs preferentially, along the grain boundaries. Moreover, the diffusivity of Cr in the austenite matrix is much lower than that of C. So, while carbon from the bulk can easily migrate to the grain boundaries, the low diffusivity of Cr causes a steep depletion of Cr around the grain boundaries (Figure 3.7). The loss in passivating element, namely Cr, at the microscopic level affects the corrosion resistance of the grain boundary area. This aspect is illustrated using the micrographs of Figure 3.8. It is clear that the grain boundaries of unsensitized Type 304SS (Figure 3.8a) are thin, like any other normal alloys, exhibiting the so called "step-like structure" across the grain boundaries, while the same alloy exhibits dark lines called "ditch type attack" when it was sensitized through heat treatment (Figure 3.8b). The ditch-type attack is due to intergranular corrosion suffered by the alloy during etching.

3.6.1.2 Kinetics of Sensitization

Chromium carbide precipitation, like any other solid state transformation, depends not only on the temperature, but also

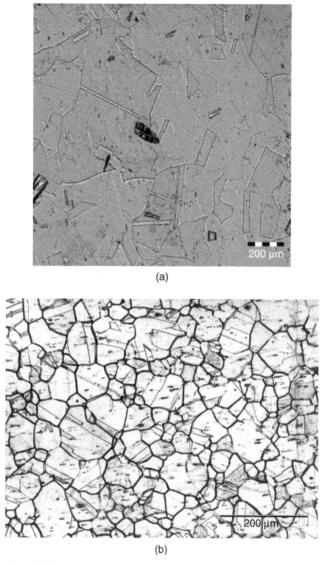

Figure 3.8 (a) Typical step-structure and (b) ditch structure of Type 304 SS in the solution-ized and sensitized conditions, respectively.

on time of exposure at the temperature. Time-temperature-sensitization (TTS) diagrams, Figure 3.9, show the relationship between temperature and the time for initiation of IGC.

The following characteristics of sensitization can be seen from TTS diagrams.

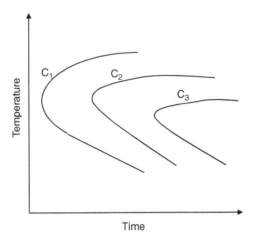

Figure 3.9 Schematic of Time–temperature-sensitization diagram showing shift in the nose of the curves towards right with decreasing carbon content. The carbon content decreases in the order C1 > C2 > C3.

1. With temperature, tendency for sensitization goes through a maximum. At low temperatures the precipitation of $Cr_{23}C_6$ is hampered by sluggish Cr diffusion and at high temperatures the poor nucleation kinetics of $Cr_{23}C_6$ controls the sensitization.
2. The curve moves towards right side, indicating lower rate of sensitization, as the carbon content of alloy is lowered. A practical application of this concept in alloy design is in the development of low carbon stainless steels such as Types 304L SS and 316L SS, which exhibit higher resistance to sensitization than Types 304 SS and 316 SS, respectively.

3.6.1.3 Factors Controlling Sensitization Kinetics The following are the major three ways by which sensitization kinetics can be lowered.

1. Lowering the carbon content.
2. Adding carbon getters (such as Ti, Nb, Ta) to fix the carbon
3. Solution annealing heat treatment.

Lowering carbon levels in austenitic stainless steels has resulted in a variety of low carbon stainless steels namely, Types 304L, 316L, and 904L SS, and adding getters as mentioned above lead to stabilized grade stainless steels such as Types 321 and 347 SS. This aspect will be discussed more in detail in Chapter 4.

Duplex stainless steels, which are being increasingly used in chemical process industries (CPI), are more resistant to sensitization, while the ferritic and austenitic stainless steels are prone to this problem. The presence of both austenite and ferrite phases in stainless steels has been found to be beneficial due to the following reason. Higher Cr diffusivity in ferrite than in austenite enables preferential growth of carbide within the ferrite phase, with shallow Cr depletion profile ahead of the carbide phase. So the ferrite phase offers more resistance to IGC. On the austenite front, though Cr depletion is severe, IGC is limited to very narrow region, which soon gets eliminated.

3.6.1.4 Environmental Effects Mere sensitization of stainless steel does not cause IGC in all types of environments. Selectivity of a sensitized alloy to an environment to suffer IGC depends on electrochemical characteristics the metal/environment exhibits. The depletion of Cr surrounding the grain boundary can promote accelerated attack provided the corrosion potential lies either in the active or in the transpassive region of the sensitized alloy. Accordingly, any environment that exhibits either a strongly reducing or strongly oxidizing behavior can cause IGC. In addition, those ions that readily destabilize the passivity of stainless steel are also responsible for IGC. Fontana (1987) provides a list of environments those cause IGC of sensitized stainless steels.

3.6.2 Weld Decay of Stainless Steels

Sensitization has been a cause of concern in austenitic stainless steels, especially in Type 304 SS weldments. Stainless steel weldments suffer intergranular corrosion at heat-affected zone (HAZ), a long strip region of the base alloy located away from the weld fusion zone as well as lying parallel to the latter (Figure 3.10).

HAZ suffers IGC because of sensitization only at this region. The fusion zone, despite experiencing higher temperature than that experienced by HAZ, does not suffer from IGC. According to the TTS diagram, a stainless steel is sensitized provided it follows the two conditions, namely,

1. Exposed to sensitization temperature regime.
2. Held for a time longer than the critical time required for sensitization. This is related to the location of the nose of the curve with respect to the time axis (see TTS diagram shown in Figure 3.9).

Among the fusion zone (FZ), HAZ and base metal (BM), only the HAZ meets both the criteria and so gets sensitized.

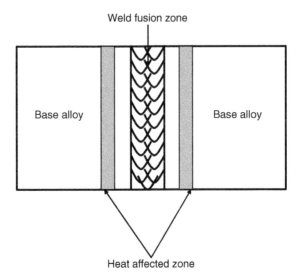

Weld fusion zone

Base alloy

Base alloy

Heat affected zone

Figure 3.10 Schematic of weldment showing weld-fusion zone, heat-affected zone, and base alloy. The long strip region of the base alloy located away on both the sides of the weld fusion zone, as well as lying parallel to it, suffers IGC.

3.6.2.1 *Parameters Controlling Weld Decay* The parameters that can influence weld decay are

(a) Those factors affecting cooling rate of HAZ
(b) Weld design
(c) Alloy chemistry (was discussed earlier)

In general, weld-cooling rate can be related to weld heat-input, which is governed by welding technique, thickness of the sheet to be welded, welding speed, arc voltage and current, and the nature of the weld joints. Among the conventional weld joining techniques, arc welding provides the lowest heat input, while the non-conventional laser and electron beam welding techniques give even lower heat input to get a sound weld than arc welding. As the heat input becomes smaller, the chances of sensitization of HAZ become minimized, so such welds exhibit high resistance to weld decay. Thick plates require large heat-input. Hence, the sensitivity of the alloy to weld decay can increase with thickness. In order to avoid weld decay in thick plates, multi-pass welding is practiced. Here the heat is allowed to dissipate after successive weld-passes. Even during multipass welding enough time-gap between each weld passes must be maintained so as to bring the temperature of the weldment to permitted levels. In fabricating industrial components weld geometry can play a crucial role in either enhancing or stifling heat transfer. It is essential that this factor is taken into account. This happens especially when thick sections are welded to a relatively thinner section of a component. While thick section dissipates the heat effectively, the thinner section can accumulate heat, due to poor heat transfer, and thereby cause weld decay.

3.7 SELECTIVE DISSOLUTION/SELECTIVE ATTACK

Alloys made of elements highly differing in electrochemical potentials have a tendency to suffer through the less noble element leaching out of the alloy, on exposure to corrosive environments. As a consequence they lose their mechanical strength leading to premature failure of components they are made with. Premature failure of heat exchanger tubes of Cu–Zn and Cu–Ni alloys and underground cast iron pipelines are common examples of this type of failure. In Cu–Zn and Cu–Ni alloys, Zn and Ni are, respectively, removed and are being called dezincification and denickelification, respectively. In cast iron Fe dissolves selectively and is called graphitic corrosion or graphitization.

3.7.1 Characteristics of Selective Dissolution

(a) Higher the electrochemical potential difference between the alloying elements, larger is the tendency for selective dissolution

(b) Increase in the content of the noble element decreases the selective dissolution of active element

Accordingly, it can be said that Cu–Ni alloys exhibit higher resistance to selective dissolution than Cu–Zn alloys and in the brass family 15% Zn does not undergo selective dissolution, while yellow (α) brasses (with about 28% Zn) are prone to selective leaching and $\alpha + \beta$ and β brasses having 40% and above Zn are even more prone to dezincification than yellow brasses.

Two types of selective leaching are industrially very important. They are dezincification and graphitic corrosion. They are discussed below.

3.7.2 Dezincification

Several mechanisms are proposed (Kaiser 1987). The most popular and simple mechanism, applied for brasses is given below.

$$CuZn \rightarrow Cu^{2+} + Zn^{2+} + 4e \qquad \text{Step I} \qquad (3.1)$$
$$CuZn + Cu^{2+} \rightarrow Cu(Cu) + Zn^{2+} \qquad \text{Step II} \qquad (3.2)$$

In the first step, both Cu and Zn dissolve together and in the subsequent step, Cu^{2+} deposits back on the surface displacing Zn from the alloy.

3.7.2.1 Nature of Attack Brasses suffered by this type of corrosion are characterized by the change in color from yellow of the brass to red, the latter being that of pure Cu. The dezincified surfaces can also exhibit other features, depending on the type of attack the alloy suffered. Two types of attack, namely, (a) uniform or layered and (b) plug type are known to occur. Figure 3.11 shows the latter type of attack found in Cu–Zn alloys (Verink and Heidersbach 1972).

High Zn content of the brass and more acidic pH of the environment favor the first type of attack, while low Zn content and neutral, alkaline, and slightly acid environments promote the latter type of attack. In addition, chlorine severely affects dezincification of alpha brasses.

3.7.2.2 Controlling Dezincification Dezincification of alpha brasses, having about 30 wt.% Zn, can be controlled by adding 1 wt.% Sn. In addition, elements such as arsenic, antimony and phosphorus in the range 0.02–0.06 wt.% are added to alpha brasses, as they are very effective in lowering dezincification. Such brasses are called Admiralty brasses. It has been found that these elements are not effective in preventing dezincification of $\alpha + \beta$ and β brasses whose zinc content lies around 40%.

When dezincification becomes a severe problem, particularly in saline water applications, then cupronickel (Cu–Ni) alloys are chosen for

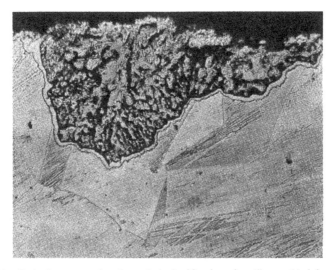

Figure 3.11 Etched cross-section through dezincification plug (Source: Verink and Heidersbach (1972). Reprinted with permission from STP 516- Localised Corrosion- Cause for Metal Failure Copyright ASTM International.).

application. Nickel being nobler than zinc, its ability to leach out from the copper alloy is slower than that of zinc.

3.7.3 Graphitic Corrosion

This type of corrosion is associated with gray cast iron. Gray cast iron pipelines, due to low cost, were largely in use for transportation of chemicals and water. The underground pipelines were found to suffer from selective dissolution of iron. As shown by the micrographs of Figure 3.12a, gray cast iron consists of graphite flakes embedded in a ferrite matrix.

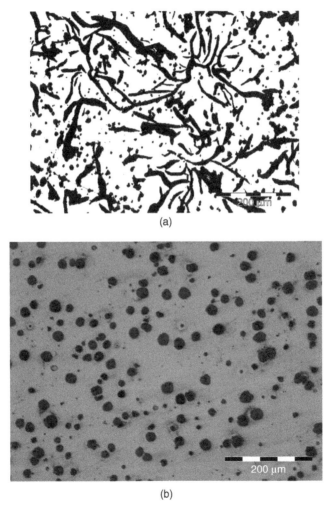

(a)

(b)

Figure 3.12 (a) Microstructure of gray cast iron showing the graphite flakes in ferrite matrix, (b) Microstructure of ductile iron showing nodular graphite. (Source: Raja (2008). Reproduced with permission of Narosa Publishing House.)

Electrochemically graphite is noble than iron and so the former induces the dissolution of the latter. Though each of the dark graphite flakes appears to be isolated from the other, the flakes in fact form a graphite network along the three directions of space. In cross-sectional images, they look separated. Such an arrangement of microstructure is very conducive for a sustained selective attack within the alloy at microscopic level. Gray cast iron pipelines suffered from graphitic corrosion; lose their strength and ability to withstand impact loading though they might appear to be intact. Soils of low pH and high salt contents cause high graphitic corrosion.

Graphitic corrosion can be prevented by employing spheroidal cast iron instead of gray cast iron. These pipelines are popularly known as ductile iron (DI) pipelines. In this type of cast iron, graphite is present as isolated spherical particles (Figure 3.12b). Notably even in such a case graphite particles on the surface exert galvanic corrosion. However, they are detached from the surface, once some amount of iron surrounding the particle dissolves. The galvanic corrosion ceases to operate once the graphite particles are detached from the surface and so DI pipes are resistant to graphitic corrosion.

3.8   FLOW-ASSISTED/EROSION/CAVITATION CORROSION

Relative motion of corrosive fluid flow with respect to structural components can induce accelerated corrosion. Fluid flow in several industrial equipments such as heat exchangers, pumps steam generators, hydraulic systems, pipelines, and nozzles suffer such corrosion. Depending upon the flow conditions and nature of environments, the mechanism of attack differs. Broadly, alloys fail by three types of flow-accelerated corrosion mechanisms:

   A. Flow-assisted corrosion
   B. Erosion corrosion and
   C. Cavitations damage

For detailed understanding of the mechanisms and the factors affecting flow-accelerated corrosion, the readers can refer very good articles by Weber (1992) Postlethwaite and Nesic (2011). A brief account of these three types of failures is made here.

3.8.1   Flow-Assisted Corrosion (FAC)

In this form of corrosion, the main role of fluid flow is to enhance mass transport (removal of corrosion product from the surface and transport of reactants to the surface) and lower protective film forming tendency of metals and environments. These are assisted by turbulent flow conditions. The mechanical damage of protective film due to fluid flow is, however, not significant although film damage by fatigue loading suggested one of the causes of failures. Flow-assisted failures are widely reported to occur in carbon steel primary water circuits of light water reactors of nuclear power plants and economizers of thermal power plant which deal with high pure deaerated water. The absence of oxygen lowers the tendency of steel to passivate in pure water. Irrespective of the flow velocity, flow-assisted corrosion in such cases found to be specific to operating temperatures of the waterlines. At high and low temperatures the carbon steel pipelines suffer low corrosion rate by this mechanism. High temperatures and relatively high oxygen levels promote protective oxide film formation, lowering the corrosion rates. Flow-assisted corrosion is characterized by "Tiger Strips" features as illustrated by Figure 3.13 (Kain et al. 2011).

Flow-assisted corrosion thus can be controlled by:-

(a) Increasing the oxidising (potential) power of water by introducing limited extent of O_2 and/or
(b) Applying alloy steels with chromium content.

3.8.2   Erosion Corrosion

At high velocities and corrosive conditions, components suffer corrosion and mechanical damage. The mechanical impact assists corrosion as well as physical removal of materials. Localized "turbulence" and "impingement" are manifestations of the velocity effect. These two phenomena increase the metal deterioration several fold at localized places when they occur. The attack can be further aggravated when two-phase fluid flow involving any of the three, viz., solid, liquid, and gas, occurs. Thus, if the corrosives such as water, acids, and gases contain solid particles such as sand and oxides, either the operating fluid velocity of pipe lines needs to be significantly lowered or the material of construction needs to be upgraded.

Inlet-end corrosion and impingement attack are two variations of erosion corrosion, which are location-specific and are widely reported. Inlet-end erosion usually observed at the inlet ends of condenser and heat exchanger tubes is a typical example of the turbulence effect (Figure 3.14a). Similarly, "impingement" attack occurs when fluid strikes metal surface head-on, like shell side fluid striking the tube outside surfaces in a shell and tube heat exchanger.

It is a noteworthy fact that high velocities can also be actually beneficial in avoiding localized under-deposit corrosion phenomena such as pitting and

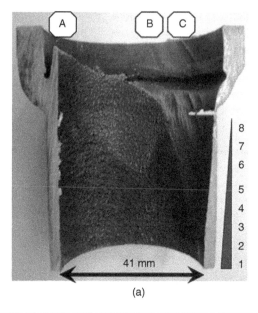

(a)

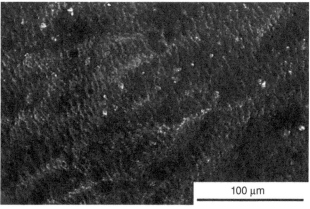

(b)

Figure 3.13 (a) Heavy thickness reduction near the socket weld (at location 8) and lower thickness reduction as seen at a distance 40 mm away from the socket weld (location 1) due to flow-induced corrosion is shown. (b) Characteristic features (Tiger Strips) of flow-induced corrosion seen at high magnification is shown (Source: Kain et al.(2011). Reprinted with permission from Elseiver.).

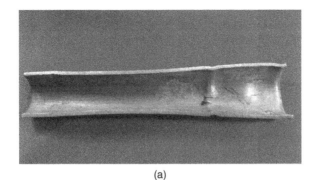

(a)

(b)

Figure 3.14 (a) 90Cu10Ni heat exchanger tube using seawater leaked within four months of operation due to erosion corrosion (b) Type 316 SS agitator blades of a reactor employing sulfonic acid rotating at a tangential velocity of 8.54 m/s failed within a few months.

crevice corrosion by preventing deposition of silts and other foreign objects. This is the reason that many stainless steels are successfully used in seawater service, like travelling water screens, provided that the sea water is kept moving at a substantial velocity.

3.8.2.1 Components More Prone to Erosion Corrosion Practically all chemical process equipments are vulnerable to erosion corrosion, but the failure due to erosion corrosion is detected very early in those equipments where the relative velocity of the moving fluid is very high. The latter can be classified into the following two categories:

1. Pipelines, particularly bends, elbows, and tees, heat exchanger tubes particularly the inlet positions, reaction vessels, distillation columns, pump casings, valve bodies and stems, etc. where the metal (the equipment) is static and the fluid moves
2. Pump impellers, blowers. Propellers, centrifuges, reaction vessel agitators (Figure 3.14b), etc. where the metal also moves

3.8.2.2 Characteristics of Erosion Corrosion Attack   This phenomenon is characterized by the appearance of grooves, waves, and valleys (Figure 3.14) that have a directional pattern along the flow direction of the corroding fluid.

Factors affecting Erosion Corrosion
   The following factors strongly influence the erosion corrosion tendency of any component.

(a) Ability of the alloys to form strong passive film
(b) Mechanical properties of alloys like strength and hardness
(c) Fluid characteristics like velocity and the content of fine solid particles (like slurry)
(d) Component design

   Now we briefly discuss the above points. Erosion corrosion resistance of an alloy can be increased by increasing the corrosion resistance of the alloy. This can be done by either enhancing passivity or increasing the inherent (thermodynamic) corrosion resistance of the alloy. It is known that copper addition to stainless steels increases the erosion corrosion resistance of the latter in sulfuric acid solutions due to the fact that copper enhances passivity of stainless steels in sulfuric acid solutions. Similarly, Cu–Ni alloys offer better resistance to erosion corrosion than Cu–Zn alloys as the former has high thermodynamic resistance to corrosion than the latter. High strength and hardness of the alloy also play important role in increasing the erosion corrosion resistance of the alloys. Thus, alloying elements (Ti, Fe) are added to Cu–30 Ni alloy to further enhance the erosion corrosion resistance, as these elements enhance hardness of the alloy.
   Change over from laminar to turbulent flow of the liquid can influence erosion corrosion resistance of the alloy. This can happen due to an increase in fluid velocity and/or deposits of the pipeline and sudden change in the direction of the flow due to component design (elbow, U-bend, etc.). It is necessary to point out the fact that the velocity effect may be nil or increase slowly until a critical velocity is reached, beyond which the attack may increase at a rapid rate. More details of the materials selection for erosion corrosion application can be found in Chapter 4.

3.8.3 Cavitation Damage

Cavitation damage primarily occurs when the hydrodynamic conditions of liquid flow favor formation and collapse of bubbles. At low pressure liquid evaporates and at high pressure, vapor condenses to form a bubble. If the external pressure becomes very high the bubble becomes unstable and implodes. The energy associated with the implosion is transferred onto the metal surface that causes not only surface oxide film removal, but also severe plastic deformation. Though the film can reform due to alloy environment interaction, it is again subjected to breakage due to bubble implosion. Such an attack causes deep pitting as brought out in Figure 3.15.

What distinguishes pitting corrosion from cavitation pit is that the former is usually covered by corrosion products, while the latter appears bright and is free from any corrosion products.

Cavitation damage is lowered by providing hard (such as stellite) or resilient (such as rubber lining) coatings.

3.9 STRESS CORROSION CRACKING

Caustic embrittlement of steel boilers and season cracking of brass cartridges are classic examples of stress corrosion cracking observed as early as 1865 and 1906, respectively. Stress corrosion cracking (SCC) continues to be an important problem for several industries, as the number of alloys–environment combinations that cause SCC has been steadily

Figure 3.15 Cavitation damage appearing as bright pits on a damaged propeller shaft of a ship. The alloy is beta brass.

increasing. For a detailed study on this topic the recent book edited by Raja and Shoji (2010) can be referred.

An alloy is said to have failed by SCC, if the cracking has occurred due to a conjoint action of tensile stress and corrosive environment. The role of stress along with the environment is one of synergy and both must act simultaneously. However, what is interesting and is beneficial from engineering perspective is that only certain alloy–environment combinations can cause SCC. Actually, the problem associated with SCC is not just the tendency of an alloy to fracture as much as the way it fractures and the stress at which it becomes susceptible to failure. The fracture is mostly brittle in nature and an alloy can fail much below its tensile strength. Hence, the failure can be rapid and can occur below the stress levels specified by the codes. The design engineers, therefore, find it difficult to design an engineering component that is expected to suffer SCC. Because of this, SCC is of industrial importance and is widely studied.

3.9.1   Characteristics of SCC

SCC process constitutes initiation and growth of cracks in an alloy, though the final failure of a component always happens to be due to mere overload mechanical failure.

Stress corrosion cracks are brittle in nature and they exhibit crack branching (Figure 3.16). Cracks almost travel perpendicular to the stress axis. Microscopically, the cracks can grow either transgranularly or intergranularly. Typical transgranular stress corrosion cracks (TGSCC) of Type 304 SS and intergranular brittle fracture of 7010 Al-alloy observed on the fractured surface lying perpendicular to the stress axis are shown in Figure 3.17 for illustration.

The type of cracking an alloy might suffer seems to depend on the metallurgy of the alloy and the nature of environment. Carbon steels in several environments are known to undergo intergranular mode of cracking. The solution-treated austenitic stainless steels crack transgranularly in chloride medium, but on sensitization the cracking mode changes to intergranular mode. High strength aluminum alloys suffer intergranular mode of cracking in chloride containing environments. Brass can suffer either transgranular or intergranular fracture depending on the pH of the environment.

Broadly speaking, grain boundary chemistry and dislocation structure influence the nature of cracking. Intergranular cracking in carbon steels, in sensitized austenite stainless steels, and in high strength aluminum alloys is caused due to higher electrochemical reactivity of the grain boundary and/or the grain boundary area than the interior of the grains of these alloys. On the other hand, the high stacking fault energy of dislocation and local ordering

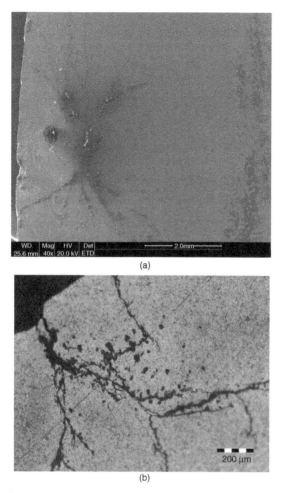

(a)

(b)

Figure 3.16 (a) Stress corrosion cracks of type 304 L SS tube of a high-pressure heater of a boiler of a thermal power plant. The cracking occurred under a deposit. Note the SS tube was free from any corrosion except on the cracked region. (b) The cross-sectional image of the cracked area of the tube showed branching cracks, typical of stress corrosion cracking. (Source: Raja (2008). Reproduced with permission of Narosa Publishing House.).

cause transgranular cracking in austenitic stainless steels. These aspects will be further discussed under SCC mechanisms.

3.9.2   Effect of SCC on Mechanical Properties

Industrial components fail prematurely, if the conditions, viz., environment, alloy metallurgy, temperature, and tensile stresses, favor SCC. The tensile stress referred here would encompass both applied and residual stress. The residual stress can arise out of any of the following factors: fabrication

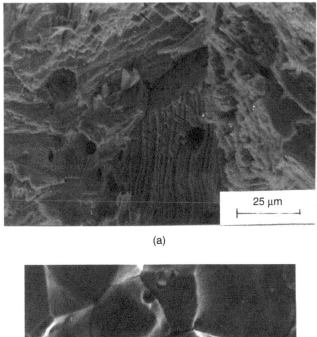

(a)

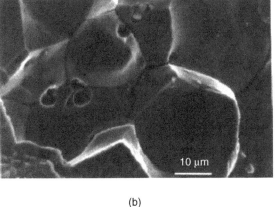

(b)

Figure 3.17 Fractographs of type 316L and peak aged 7010 Al-alloy subjected to SCC tests through U-bend sample showing transgranular and intergranular cracking, respectively. Type 316L was subjected to boiling 42% $MgCl_2$, while 7010 Al-alloy was subjected to 3.5% NaCl at room temperature. (Source: Raja (2008). Reproduced with permission of Narosa Publishing House.).

process, such as welding, grinding, machining, rolling, etc. In addition, the accumulation of corrosion products within the wedges of a component can also exert tensile stresses. The presence of both of these stresses lowers the allowable operating stress levels.

SCC can severely affect allowable operating stresses and fracture toughness. Notably, failure can occur even if the components experience stress levels much below the yield strength of the alloy. Cases of components failure, even if they were subjected to 20% of their yield strength, are known.

Stress corrosion cracking is a time-dependent process. For a given environmental conditions the time to failure depends on the applied stress. The rise in stress lowers the time to failure (Figure 3.18) (Denhard Jr. 1960). One of the characteristics of SCC phenomenon is that the alloys undergoing SCC exhibit threshold stress below which the alloys become resistant to SCC as has been shown by the above Figure.

3.9.3 Factors Affecting SCC

The factors affecting SCC can be broadly classified into two types, those concerning the metal/alloy and the others related to the environment. However, the interaction of the environment with the metal produces another factor, which is electrochemistry. The following issues are the major concern for SCC resistance of an alloy

(a) Nature of environment
(b) Passive film stability
(c) Metallurgy
    i) Phases
    ii) Grain boundaries
    iii) Dislocations

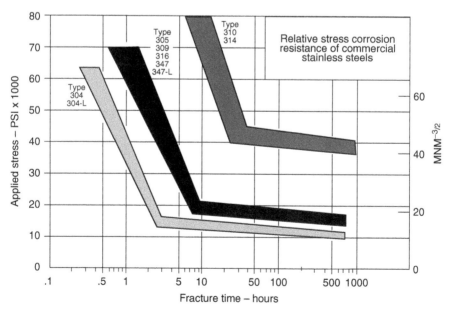

Figure 3.18 Composite curves illustrating the relative stress corrosion cracking resistance for commercial stainless steels in boiling magnesium chloride. (© NACE International 1960.)

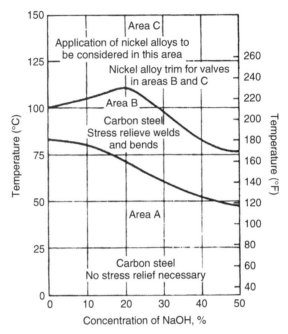

Figure 3.19 Carbon steels: Temperature and concentration limits for stress-corrosion cracking in NaOH. (© NACE International 1974.)

iv) Grain size and orientation

These aspects are briefly discussed below.

3.9.3.1 *Nature of Environment* It is known that only certain metal/environment combination can cause SCC, though the list of environment causing SCC is expanding. It is also necessary to point out the fact that the definition of environment could mean nature and concentration of the chemical species, conditions of stagnancy or flow, and the temperature of the environment. Hence, SCC failure avoidance hinges on the right choice of material based on proper assessment of industrial environment on one hand and maintaining the environment parameters within acceptable levels for a chosen material on the other. These aspects are illustrated below.

Carbon steels are known to suffer SCC in caustic environment, but an examination of Figure 3.19 shows that within certain caustic levels and operating temperatures the steel can be resistant to SCC even if it possesses residual stresses (Corrosion Data Survey 1985). Should the temperature be raised, the carbon steel component should be stress relieved, without which it will suffer SCC. From the diagram, it is also clear that beyond a temperature limit only nickel base alloys or nickel can be employed and stress-relieving carbon steel does not help to prevent SCC.

Successful application of stainless steels for various industrial processes again depends on examining the limiting conditions that could cause SCC. Now it is well known that austenitic stainless steels can be used for cooling water systems only if the cooling water is maintained below 50 °C, above which they are prone to SCC. Should the temperature exceed either by design or by poor operating practices, either Cu-base alloys or duplex stainless steels need to be selected. The successful application of stainless steels in boiler water situations, whose operating temperature far exceeds the above limit, relies solely on controlling the O_2 and Cl^- levels of boiler water.

It is also recognized that in condensers and coolers, made of stainless steel tubes, the tendency of cooling water (with the same chemistry) to cause SCC in the tubes much depends on the path of the cooling water and orientation of the heat exchangers. The coolers exhibit more resistance to SCC when water flows through tube side than through the shell side. High stagnancy of water in shell side is responsible for larger incidence of SCC failures of stainless steel tubes of vertical heat exchanger.

3.9.3.2 Passive Film Stability Passivity is a thermodynamically unstable state of an alloy, though the corrosion resistance of a metal/alloy is achieved through passive film formation. Potential-pH (Pourbaix) diagrams are meant to delineate potential-pH regimes of immunity to corrosion, tendency for dissolution, and passivity of a metal. SCC is known to occur within the passive regime of an alloy. Hence, these diagrams are used to highlight regions of SCC susceptibility of various alloys (see Figure 2.4). Notably, carbon steel passivates in nitrates, carbonates, caustic, and phosphates environments. So these environments induce SCC in steel. As these anions passivate steel in different potential–pH range, they inflict SCC on steel at different potential–pH conditions.

Initiation as well as the propagation of SCC in an alloy much depends on the passive film stability. This is especially the case, if the SCC mechanism is governed by film rupture process. When the potential (corrosion potential/ applied potential) of an alloy lies in the unstable passive region, such as active–passive region of the curve (Figure 3.20), the alloy tends to crack through SCC. Chlorides are known to destabilize the passive film and hence cause SCC, a type of "calm followed by storm" situation. It is also known that pits become active sites for SCC initiation.

3.9.3.3 Metallurgy Alloy chemistry and crystal structure, nature of phases and their distribution, non-equilibrium segregation of alloying elements,

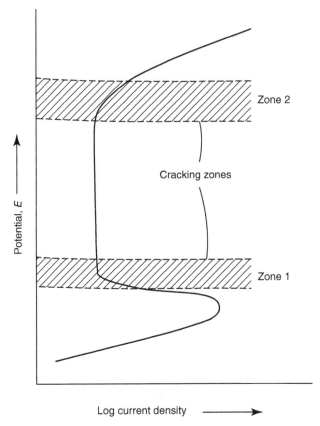

Figure 3.20   Stress corrosion cracking susceptible areas with respect to anodic polarization curve exhibiting active-passive-transpassive transition.

type of dislocation, grain size, morphology, and its orientation are some of the important metallurgical variables which influence SCC. These are briefly discussed below.

Ferritic stainless steels with body-centred cubic structure are resistant to SCC, while austenitic stainless steels, with face-centered cubic structure, are susceptible. Addition of Ni in small levels brings down the SCC resistance of stainless steels. Similar effect is noticed with respect to nitrogen as well. However, at high levels both Ni and N promote cellular dislocations and so become beneficial towards SCC resistance. Notably duplex stainless steels having both ferrite and austenite phases are found to be superior to both austenite and ferritic stainless steels in chloride environments. In carbon steels, the elements C, P, S, and N influence SCC, by modifying the grain boundary chemistry. Chapter 4 discusses the role of metallurgy, properties of alloys, and alloy selection for SCC resistance in detail.

3.9.4 Controlling SCC

Several of the following methods can be judiciously applied to control SCC

1. Use alloys which are resistant to SCC
2. Eliminate species (such as chloride and oxygen) responsible for SCC
3. Lower the tensile stress. If it is residual, consider post-weld/work heat treatment
4. Apply compressive stresses through methods such as shot-peening
5. Lower stress concentration in components
6. Apply coating
7. Apply cathodic protection

3.10 HYDROGEN DAMAGE

Hydrogen in metals/alloys such as steels, stainless steels and alloys of aluminum, titanium, magnesium, zirconium causes premature cracking and lowers ductility and toughness. The mechanism of cracking is wide ranging and depends on material, environment, loading conditions, temperature, etc. It is possible to broadly classify the hydrogen-related failures into the categories as below.

- Hydrogen cracking occurring at low temperature aqueous/wet conditions
  - Stepwise cracking (SWC) or hydrogen-induced cracking
  - Hydrogen embrittlement
- Metal damage occurring at elevated temperatures, in presence of hydrogen containing gases and/or steam.
  This latter category may or may not involve corrosion as a step.

Hydrogen atom, being the smallest of all the elements in the Periodic Table, easily diffuses into the metal lattice. Hydrogen in the lattice can cause cracking through various mechanisms giving rise to different forms of hydrogen-assisted cracking, which will be discussed in the subsequent sections.

3.10.1 Low Temperature Hydrogen-Induced Cracking

Ambient/low temperature hydrogen damage is further subdivided into the following four classes. The atomic hydrogen after diffusing from the surface to the interior of the steel can lead to the following types of failures.

(a) Hydrogen Blistering: Blister formation in low-strength steels initiated in pores/cracks and interface of nonmetallic inclusions, such as sulfides, and the matrix

(b) Hydrogen Induced Cracking (popularly known as HIC), also known as SWC (Step Wise Cracking)

(c) Sulfide Stress Cracking (SSC): cracking occurring in steels of high strength/hardness.

(d) Stress-Oriented Hydrogen-Induced Cracking (SOHIC)

A schematic illustration of the occurrence of the four types of hydrogen damage on a weldment of a steel pipeline exposed to hydrogen sulfide gas is shown in Figure 3.21 (Elboujdaini 2011).

Hydrogen Blistering (HB) and SWC/HIC are similar in the sense that the hydrogen atoms form molecular hydrogen at pores, cracks, and inclusion/matrix interface and develop internal pressure. When the pressure reaches a critical value, microcracks develop and grow. As these defects are generally planar in nature, these cracks look like steps, leading to the term SWC (Figure 3.22a). These cracks over the time period grow, join, and become large in size leading to bigger blisters (Figure 3.22b). Failure analysis by Raja and Murugesan, (1992) gives detailed information on the role of microstructure on SWC and HB of steels.

Hydrogen embrittlement (HE)/(SSC): The former term is more generic, while the latter refers to HE in sulfide containing environments. Sulfides promote hydrogen uptake of steels. In these cases, the hydrogen remains in the lattice in atomic form and promotes brittleness through several mechanisms

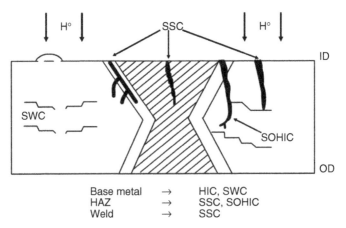

Figure 3.21 Forms of hydrogen damage in H_2S service. See the text for the explanations of the abbreviations. (Source: Elboujdaini (2011). Reprinted with permission from John Wiley & Sons, Inc.)

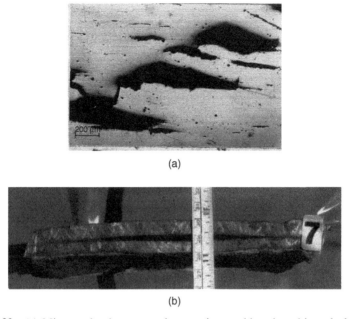

(a)

(b)

Figure 3.22 (a) Microcracks that appear in stepwise cracking that ultimately leading to blister as shown in (b).

(Lynch 2010). Notably, higher the strength and hardness of steel/alloy, higher will be its susceptibility to cracking. Steel welds are more prone to HE/SSC as they exhibit residual tensile stresses as well as high hardness. *Notably, hydrogen can also form brittle hydrides phases with metals such as Ti, Mg, and Zr and thereby promote cracking. This aspect is not discussed over here, but can be found in the above reference.*

SOHIC: In simple terms, SOHIC has mixed characters of HB and HE. So applied stresses as well as stresses developed by molecular hydrogen are responsible for cracking. Notably, SOHIC tends to occur in the base metal region adjacent to hard weldments, mostly in plate steels, where cracking might initiate by SSC. SOHIC is characterized by interlinking microscopic cracks oriented both in the direction perpendicular to that of the stress and in the plane defined by nonmetallic inclusions.

Turning back to the Figure 3.21, it is clear that the above-described failures occur at different positions of the weldment. A close examination of the weldment would show that weld-fusion zone is generally harder than the remaining area of the pipeline and is also subjected to tensile stresses. Accordingly, this region as per the above discussion must be susceptible to HE/SSC. On the other hand, the pipeline is soft, exhibits low strength, and so can be prone to SWC and HIC if the accumulated hydrogen gas develops reasonable pressure, without needing any external stresses. Both SWC and HIC will turn into blisters close to the surface, as low stresses developed by hydrogen gas will be sufficient to form blisters at the surface. The mixed character of SOHIC enables it to occur between the two regions where HE and HIC occur.

Process Variables affecting Hydrogen-related Cracking Phenomena.

- Presence of water is essential to all the above four classes of hydrogen damage, since hydrogen is produced from water through cathodic part of the corrosion reaction.
- Generally, wet hydrogen attack does not occur in neutral environments. The environment needs to be acidic (pH below 4) if only H_2S is present and even alkaline environments (pH above 8) can be effective, if dissolved cyanide is also present.
- H_2S levels above 50 ppm are normally required for the attack.
- Most of the wet hydrogen-related attack occurs at operational temperatures between ambient and about 150 °C. SSC is quite common around 80 °C.
- Presence of Cl^- and CO_2 along with H_2S, such as that occurring in oil wells, accelerates hydrogen cracking phenomena through reduction of pH and removing protective corrosion product scale.
- In addition to the above factors, pickling of steels, use of wet welding electrodes, and even simple corrosion phenomena can introduce hydrogen into the materials.
- The species such as sulfides, cyanides, arsenic, antimony, and phosphorus in the corrosive environments or on the surface of the material can enhance hydrogen permeation into the material.

3.10.1.1 Controlling Low Temperature Hydrogen-Induced Cracking
The following are some of the important suggestions for increasing the resistance of materials/component against low temperature hydrogen-induced cracking

For SWC/HIC/HB

- Reduce the proportion of nonmetallic inclusions by lowering the S content
- Modify the morphology of S segregation by adding Ca which spheroidize the nonmetallic inclusions

HE/SSC

- Lower the tensile stresses
- Providing right tempering heat treatment in the case of steel to eliminate detrimental effect of martensite in steels. In the case of aluminium alloys provide a overaging treatment.
- Prefer the presence of Cu (>0.2%) as an alloying element
- Prefer plate mill products over hot strip mill products for the steel
- Avoid hydrogen entry into steel during surface treatment and fabrication
- If the above is not possible, bake the components to remove the hydrogen.

3.10.2 High Temperature Hydrogen Damage/Decarburization

At elevated temperatures, well above $150\,°C$, both molecular and atomic hydrogen, if present on steel surface, can react with carbides leading to decarburization and methane formation. The corresponding reactions are given below.

$$C(Fe) + 2H_2 \rightarrow CH_4 \tag{3.3}$$
$$C(Fe) + 4H \rightarrow CH_4 \tag{3.4}$$

The first reaction occurs at the steel surface, while the second reaction can occur either on the surface or in the bulk of the steel. This is possible because only atomic hydrogen can diffuse into the steel. Accordingly, the steel can suffer two types of decarburization/hydrogen attack.

The presence of hydrogen environment over industrial components is sometimes intentional and in some other times unintentional. Hydrocarbon industries deal with large quantities of hydrogen at various plants/units, which is intentional. On the other hand, hydrogen generated as a result of steam/water reacting/corroding with steel in a boiler through the following reaction is unintentional.

$$3Fe + 4H_2O \rightarrow Fe_3O_4 + 4H_2 \tag{3.5}$$

The type, extent, and time for the damage depend upon the temperature, pressure, and partial pressure of hydrogen in the gas.

3.10.2.1 Nature of Attack When steel is exposed to hydrogen containing gaseous atmospheres at elevated temperatures it may suffer permanent damage. There are two main types of such damages which are as follows:

- Internal decarburization (generally intergranular): The methane formed within the steel cannot diffuse out and hence generates internal voids, which eventually coalesce to form microfissures and cracks. Features of such type of attack are fissuring, lamination, blistering and eventually internal cracking, and finally fracture. In this type of damage, the tensile strength of the steel would be reduced to half its original value and the material would be quite brittle.
- External decarburization, taking place at the surface, is qualitatively similar to the decarburization occurring during heat treatment/exposure to high temperature oxidizing gases. This makes the surface softer and weaker than the bulk that would have had only the temperature effect.

Figure 3.23 shows a micrograph of a steel that suffered severe high temperature hydrogen attack. Depletion of carbides is clearly visible in the micrograph.

3.11   STRAY CURRENT CORROSION

This kind of corrosion pertains to buried (in soil) or submerged (in water) structures. Stray currents are those that follow paths other than the intended circuit, or they may be any extraneous current in the earth. If currents of this kind enter a metal structure, like a buried pipeline, they cause localized corrosion at areas where currents have to leave again and get back to the structure of origin.

Sources of stray currents are commonly electric railways, grounded electric DC power systems, electric welders, cathodic protection systems, and electroplating plants. An example of stray current from an electric railway system in which steel rails are used for current return is shown in Figure 3.24 below (Revie and Uhlig 2008). Because of either poor bonding between rails or poor insulation of the rails to the earth, some of the return current enters the soil and finds a low resistance path such as a buried pipeline. Localized corrosion takes place at the areas marked B on the pipeline where the entered current leaves back to the earth. On the other hand, at the areas marked A on the pipeline where stray current from the tracks enter the pipeline no corrosion takes place.

Any of the following means, if employed, can lower stray current corrosion (Szeliga, 2001):

- Short the service pipe at B and the rail track at C using a low resistance metallic conductor (called drainage-bond) as illustrated in the Figure 3.24.

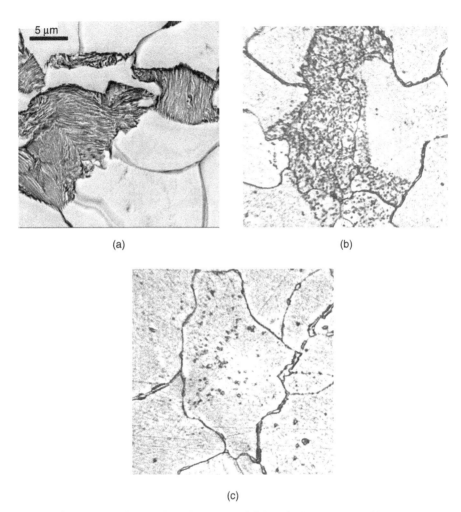

(a)   (b)

(c)

Figure 3.23 Shows micrographs of 0.1C-1.1Cr-0.26Mo-0.17V steel that suffered severe high temperature hydrogen attack. (a) unexposed and depletion of carbides after (b) 110, 000 and (c) 190,000 h of exposure to steam conditions (Courtesy: Hryhoriy Nykyforchyn).

- Apply thick coating on the stray current-prone area and extend the coating to a large segment of the pipeline on either side of the corroding area.
- Fill the corroding area with sand to provide resistance to the flow of the current in the soil.
- Install electrical shield and sacrificial anode of scrap iron at the point B. In this case, stray currents cause only the corrosion of the sacrificial anode, which is easily replaceable at low cost without any public discomfort and without any need to replace with a new pipeline.

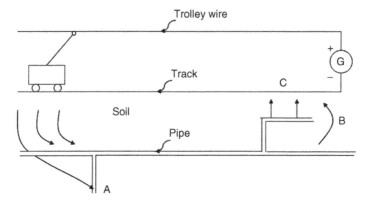

Figure 3.24 Stray-current corrosion of buried pipe (Source: Revie W and Uhlig (2008). Reprinted with permission of Wiley.).

3.12   HIGH TEMPERATURE CORROSION

Several industrial components such as boilers and gas turbines are subjected to high temperature corrosion environments resulting in severe degradation and sometime premature failures. Hence, this becomes an important subject of discussion for this chapter. However, the subject of high temperature corrosion is vast and this section by no means covers all the aspects of this problem. The recent review (Datta et al. 2008) gives a very good overview of this subject and the books by Kofstad (1988) and Young (2008) treat this subject in detail.

High temperature corrosion can be classified as

(a) Oxidation
(b) Sulfidation
(c) Hot corrosion
(d) Carburization/ metal dusting
(e) Chloridation

3.12.1   Oxidation

An idea of the thermodynamic tendency of pure metallic elements for oxidation and sulfidation at various temperatures can be obtained through the famous Ellingham and Richardson diagrams, respectively. The majority of the metals used for engineering applications exhibit large thermodynamic tendency for high temperature corrosion, although their resistance to high temperature corrosion is imparted by alloying with elements such as Cr, Al, and Si. These elements preferentially get oxidized on the metal surface to form a protective oxide scale that offers good resistance to further corrosion. Superior high temperature corrosion-resistant alloys are developed by improvising the stability of the oxide scale against spallation, lowering the cation transport in the oxide film, and enhancing both the oxygen ion migration in the oxide film and the diffusion of oxide forming elements such as Cr and Al in the alloy.

3.12.2   Sulfidation

The ability of an alloy to form a protective oxide scale becomes less under sulfidation conditions, especially when the oxygen partial pressure of the environment becomes too less. In such a case, the alloy suffers much higher high temperature corrosion than it would suffer from a simple oxidation. Hence, sulfidation-resistant alloys are designed with alloying elements such as Cr, V, Nb, Ta, Mo, and W. Coatings of the type, MCrAlY (M=Fe or Ni or Co), especially alloyed with V, Nb, Mo, or W, have been found to exhibit high resistance to sulfidation.

3.12.3   Hot Corrosion

Hot corrosion is a serious threat to the stability of gas turbine; especially those operating in marine environments. The salt, Na_2SO_4, deposited on the metal surface enhances the oxidation. This salt is formed as a result of either sulfur, present in fuel as an impurity, getting oxidized to SO_2/SO_3 and subsequently transforming to the Na_2SO_4 salt or is even present in the marine environment itself. The presence of chlorides that arise from marine environment and the V_2O_5, formed as a result of oxidation of V present in the fuel as impurity, further accelerate the hot corrosion of metals. Alumina scale is found to resist more than chromia scale. So hot corrosion-resistant alloys contain high level of aluminum or are coated with aluminides in order to enhance hot corrosion resistance.

3.12.4   Chloridation

Chlorine-containing environments can cause severe corrosion damage to metals and alloys at high temperatures. The problem is further compounded by the fact that chlorides of metals possess low melting temperatures than either oxides or sulfides, thus unable to form protective scales at high temperatures, which normally happen with oxides and sulfides. Chlorine also can damage protective oxides formed due to oxygen and make the metals less stable. Thus, the metals become more unstable in chlorine-containing environments than in oxidizing or sulfidizing gaseous conditions. Chromium and aluminum additions to steels and nickel base alloys can enhance their resistance to chlorine attack. Chromium and aluminum oxides are protective in nature and the latter is superior to the former.

The presence of chlorides in coal poses a great threat to stability of boilers. High levels of chlorides ($\geq 0.2\%$) have been found to enhance even sulfidation attack. Chlorides can also indirectly affect the boiler system, when they convert to HCl and condense on low temperature areas such as economizers.

3.12.5 Carburization/Metal Dusting

Carbon-containing environments such as those found in petrochemical, refineries, and coal conversion industries can cause carburization of steels, especially at high temperatures. The extent of carburization depends on carbon activity and temperature of the environment. Carburization leads to rapid attack called metal dusting and the surface appears as round bottom pits. The tendency of carburization is lowered by enhancing its aluminum content or aluminizing the surface altogether. From the point of taming the environment, adding sulfur (0.05–0.5 wt.%) in the form of hydrogen sulfide or mercaptan to the feed can significantly lower carburization.

REFERENCES

Bengough, D., Jones R. M. and Pirret R., (1920), J. Inst. Metals, 23, p. 65.

Corrosion Data Survey, (1985), *Metal Selection*, 6th Edition, NACE International, Houston, TX, p. 34.

Datta, P. K., Du, H. L. and Burnell_Gary, J. S., (2008), High Temperature Corrosion in *Corrosion Science and Technology: Mechanism, Mitigation and Monitoring* (Eds) U. K. Mudali and B. Raj, Narosa Publishing House, New Delhi, 2008, pp. 50–94.

Denhard Jr, E. E., (1960), Effect of Composition and Heat Treatment on Stress Corrosion Cracking of Stainless Steels, Corrosion, 16, p. 135.

During, E. D. D., (1988), Corrosion Atlas-A Collection of Illustrated Case Histories, 1, Elsevier, B-070.

Elboujdaini M., (2011), Hydrogen Induced Cracking and Sulfide Stress Cracking in *Uhlig's Corrosion Handbook*, 2nd Edition, (Ed) R. Winston Revie, John Wiley & Sons, Inc., NJ, p. 183.

Fontana, M. G., (1986), *Corrosion Engineering*. McGraw Hill, NY, New York, p. 82.

Kain, V., Roychowdhury, S., Ahmedabadi, P., Barua, D. K., (2011), Engineering Failure Analysis, 18, pp. 2028–2041.

Kaiser, H., (1987), *Alloy Dissolution in Corrosion Mechanisms*, F. Mansfeld (Ed), Marcel Dekker, New York, p. 85.

Kofstad, P., (1988), *High Temperature Corrosion*, Elsevier Applied Science, London.

Lynch, S., (2010), Hydrogen embrittlement in *Stress Corrosion Cracking*: Theory and Practice, Eds. V.S. Raja, T. Shoji and S. Lynch, Woodhead Publications, UK.

Postlethwaite, J. and Nesic, S., (2011), Erosion-corrosion in single-and multiphase flow, in *Uhlig's Corrosion Handbook*, Ed. R.W. Revie, John Wiley & Sons, Inc., New Jersey, p. 215.

Raja, V.S. and Murugesan, A., (1992), Failure Investigation of a Pressure Vessel in a Catalytic Reformer Unit, Materials Performance, 41, (1) 63–67.

Raja, V. S, (2008), *Localized Corrosion in Corrosion Science and Technology: Mechanism, Mitigation and Monitoring* U.K. Mudali and B. Raj, Narosa Publishing House, New Delhi, 1–49.

Raja, V. S, and Shoji, T., (2010), *Stress Corrosion Cracking*: Theory and Practice, Woodhead Publications, UK.

Revie, R.W., and Uhlig, H. H., (2008), *Corrosion and Corrosion Control*, 4th Edition, New York, John Wiley & Sons, Inc., p.242.

Szeliga, M. J., (2001), *Stray Current Corrosion in "Peabody's Control of Pipeline Corrosion"*, Ed. R. L. Bianchetti, NACE International, Houston, 2nd Edition, p.211.

Tuthill, A. H. and Avery, R. E., (1992), Advanced Materials & Processes, 142 (2), p. 34.

Verink E. D., Heidersbach R., 1972, Evaluation of the Tendency for Dealloying in Metal Systems, in Localized Corrosion of Metal failure, ASTM Special Technical Publication 516-EB.

Weber, J., (1992), Flow Induced Corrosion: 25 Years of Industrial Experience, British Corrosion Journal, 27, p. 193–199.

Young, D., (Ed.) (2008), *High Temperature Oxidation and Corrosion of Metals Elsevier*, London.

4

MATERIALS OF CONSTRUCTION FOR CHEMICAL PROCESS INDUSTRIES

4.1 INTRODUCTION

Primary considerations for selecting a material for manufacturing a component are governed more often by such properties as fabricability, machinability, strength, thermal conductivity, formability, hardenability, and availability. Their corrosion resistance characteristics are considered much later and it is possible that a material chosen even after due consideration of corrosion behavior may not find its end use should the cost be high. So, the corrosion engineer faces a difficult task of convincing the management on using appropriate materials for application from corrosion stand point. Of course, the issue of corrosion will be given larger importance with regard to material of construction (MOC) for chemical process industries than in other industries. Nevertheless, corrosion engineers could face stiff resistance from the project/design personnel, if not adequately justified on the selection criteria. A thorough analysis of available materials and their corrosion behavior is essential to make an appropriate choice. This chapter is meant to provide guidelines on the selection of materials, primarily metallic, for equipments in chemical process industry.

Detailed corrosion data on various materials can be obtained from several handbooks (De Renzo (1985), Craig (1989), McEvily (1990), Schweitzer (1995), Wilks (2001), Kreysa and Schutze (2005). While most of the data

Corrosion Failures: Theory, Case Studies, and Solutions, First Edition. K. Elayaperumal and V. S. Raja.
© 2015 John Wiley & Sons, Inc. Published 2015 by John Wiley & Sons, Inc.

books on corrosion provide information on uniform corrosion in various environments, the 16 volumes handbook by Kreysa and Schutze are exhaustive and the handbook by De Reno also provides information related to intergranular corrosion, pitting corrosion, and stress corrosion and the atlas by McEvily provides data related to SCC and corrosion fatigue. The following points, however, need to be emphasized when deciding a MOC.

1. Material selection need not be the only solution to the problem.
2. Corrosion resistance of a material can be supplemented by use of other mitigation techniques such as coating, inhibitors, cathodic protection, design, monitoring, and inspection.
3. Though corrosion resistance of a material depends to a large extent on the environment (chemistry, temperature, and velocity mainly) to which it is exposed to, due considerations must also be given to design and fabrication as the latter do influence the corrosion resistance.
4. While the positive characteristics of the materials, which make them suitable for the slated service, may be known, their negative characteristics, against which precautionary measures needed to be taken, are not very much known. This aspect should also be considered.

Table 4.1 lists the important groups of materials that are used in chemical process industry and gives an indication of their most common applications. These materials are those that are used for major equipments such as pressure vessels, heat exchangers, piping, etc. While these are important choice of materials for construction, other materials such as aluminum, ceramics, and plastics find increasing applications. A brief review of the properties of these latter materials is also presented in this chapter.

4.2 CAST IRONS

Cast irons are the least expensive among all engineering alloys that belong to ferrous group of materials. There are several types of cast irons, exhibiting different mechanical properties and microstructures. White cast iron, gray cast iron, malleable cast iron, and spheroidal/ductile cast iron are the major classification of cast irons. Un-alloyed cast irons are normally employed. However, they can also be alloyed in order to enhance corrosion resistance. Silicon, nickel, chromium, copper, and molybdenum have been found to improve the corrosion resistance of cast irons. High silicon cast iron (12–18 wt.%) has been known to exhibit excellent erosion corrosion resistance.

TABLE 4.1 Typical Applications of Metallic Materials of Construction for Equipments in Chemical Process Industry

Material	Usage/Application
Cast Irons	Pumps, compressors, and valves
Carbon steels	General purpose where corrosion resistance is of secondary importance
Alloy steels, including low alloy C–Mo and Cr–Mo steels	Elevated temperature applications, steam service, hydrogen service, high temperature sulfide service
Ferritic and martensitic stainless steels, including 12% Cr and 17% Cr	For resistance to sulfur-related attacks, valve trims. Pumps wear rings and heat exchanger tubings for mild corrosive service
Austenitic Cr–Ni corrosion-resistant stainless steels	All corrosion-resistant duties where corrosion is not very severe
Austenitic Cr–Ni heat-resistant stainless steels	Furnace tubes, fittings, burners, flare tips, liners/shrouds
Duplex Cr-low Ni stainless steels	For corrosive applications where stress corrosion cracking by chloride is of primary importance and for applications where corrosion resistance, erosion resistance, and mechanical strength are of equal importance
Nickel base alloys, Ni–Mo, Ni–Cr–Fe–Mo	For highly corrosive applications in chemical environments and high temperature corrosive applications
Copper	Electrical applications, tubings for steam tracings, instrument tubings
Brass	Tubes and tube sheets for coolers and condensers, instrument fittings and tubes
Bronze	Tube sheets, Valves, and other parts exposed to sea water
Aluminum brass	Tubes for sea water condensers where corrosion resistance and effective cooling are of prime importance
Cupro Nickel (both 90/10 and 70/30)	Tubes for sea water condenser service where resistance to general corrosion, dezincification and erosion corrosion, and also mechanical strength are of equal importance
Aluminum	Insulation sheathing, DM water lines, and for process equipments of oxidizing environments without chlorides
Titanium	Tubes and tube sheets for sea water cooling where lightness and strength are of equal importance along with corrosion resistance and also for certain highly corrosive applications involving acidic chlorides
Tantalum	For ultimate corrosion resistance where normally glass lining is specified

In terms of corrosion resistance, they behave in a manner similar to carbon steels. But what distinguishes cast irons from that of steels is the presence of free carbon, except that of white cast iron, which can cause microgalvanic corrosion. In fact, gray cast iron suffers from graphitic corrosion due to the presence of graphite network. As such a network is absent in malleable and ductile cast irons, these latter resist graphitic corrosion. Cast irons find application in as cast pump components, water-pipelines, steam lines, and tanks for storing acids, which generally do not involve extensive fabrication. In fact, poor weldability and toughness limit their applications. Steels are superior in this context and so find wider applications as MOC.

4.3 CARBON STEELS

Carbon steel is used in chemical process industries to a much greater extent than any other material. It is used in two main classes: Structure purpose and Process Equipment purpose. Structural steel is for pipe racks, support columns, cross beams, pressure vessel skirts, etc. where corrosion is somewhat tolerated and/or controlled by protective paint coatings. This book is not concerned specifically with structural steel. Process equipment steels are used for making items such as pressure vessels, reactors, distillation columns, trays, piping, heat exchangers, etc. where corrosion is monitored and controlled during process. ASTM Standards are widely used in construction of process plant equipments. Table 4.2 gives the ASTM Designations for carbon steels that are commonly used in process plants.

The major specific differences in the content of alloying elements such as carbon, silicon, manganese, sulfur, and phosphorus in the above specifications are intended for improving the mechanical properties such as strength, ductility, and toughness and also the weldability. As far as corrosion failures are concerned, all the above forms and specific types within the particular form behave in a comparable way. In other words, compositional variations within carbon steel do not alter the behavior against common forms of corrosion such as uniform corrosion, pitting corrosion, crevice corrosion, and galvanic corrosion—which are electrochemically controlled, not composition-centered. The exemptions are the resistance to stress corrosion cracking (SCC), corrosion fatigue, erosion corrosion and hydrogen-related cracking, including sulfide stress cracking (SSC), which are indirectly dependent upon strength and hardness characteristics of the steel. It is to be noted that strength and hardness characteristics of steels, for that matter most of the alloys, depend on heat treatment and microstructure.

The most important failure modes for carbon steel, related to corrosion, are uniform corrosion, erosion/flow-assisted corrosion, and environmentally assisted cracking (EAC). These are discussed below:

TABLE 4.2 ASTM Standards for Carbon Steels for Chemical Process Equipments

Form	Specific Type	ASTM Standard
Plate	Structural quality	A 283
	Notch ductile plate for tanks	A 131
	Boiler plate	A 285
	Impact-tested plate for shells of equipments	A 516
Pipe	Line pipe	API 5L
	Pipe	A 53
	Silicon-killed seamless pipe	A 106
	Boiler tube, seamless	A 192
	Resistance-welded	A 135
	Fusion-welded	A 134
	Fusion-welded for high temperature service	A 155
	Impact-tested seamless pipe	A 333
Fittings	Welded	A 234
	Impact-tested seamless	A 420
Castings	Cast carbon steel for general applications	A 27
	Cast carbon steel for high temperature service	A 216
Forgings	For general service	A 181
	For high temperature service	A 105
Heat exchanger tubes	Seamless cold drawn	A 179
	ERW cold drawn	A 214

4.3.1 Corrosion

Corrosion is a primary cause of failure in carbon steel equipment. Carbon steel is used for general purpose in neutral (such as water lines), in alkaline (such as boiler circuits), and in somewhat acidic environments. The mild acidic environments are generated in aqueous solutions containing CO_2, H_2S, NaCl, phenol, and other similar substances. Corrosion by such environments is accentuated by high fluid velocity and turbulence. The latter damage the protective corrosion product film and sustain the damage function. This is controlled by design and, if feasible and permissible, by addition of corrosion inhibitors and raising the environment pH. For piping and heat exchanger tubes, certain velocity limits, both minimum and maximum, are specified. The upper limit is for avoiding erosion corrosion and the lower limit is to avoid under-deposit pitting corrosion in low velocity stagnant areas. Within the operative velocities, inhibitor additions, controlling the operating parameters and also proper monitoring followed by timely replacement, are the practiced methods for corrosion prevention.

4.3.2   Stress Corrosion Cracking, Including Hydrogen Cracking and Sulfide Stress Cracking

Stress corrosion cracking is also a common mode of failure occurrence in carbon steels, but not as common and frequent as general corrosion. Stress corrosion cracking is caused by a number of different agents such as caustic alkali, cyanides, hydrogen sulfide, carbonates, and nitrates, out of which caustic alkali and hydrogen sulfide are very common and frequent. Unlike stainless steels, chloride SCC does not occur in carbon steels.

Hydrogen sulfide in chemical process industry environments may be present both in process streams and in the general atmosphere of the plants. The latter would dissolve in the condensed moisture and initiate corrosive actions on the outside bare surfaces of carbon steel equipments. Susceptibility of carbon steels to SSC and Hydrogen-Induced Cracking (HIC) increases with H_2S concentration, hardness, applied or residual stress, and decrease in pH towards acidic values. Very small amounts of H_2S can be hazardous. SSC/HIC in carbon steel is normally prevented by keeping the hardness of the steel below HRC 22 (250HV) for wrought parent metal and below HRC 20 (210HV) for welds. Post weld heat treatment (PWHT) is specified for pressure vessels which would process aqueous solutions of H_2S. The latter, in the absence of moisture, is not corrosive at ambient temperatures and somewhat higher temperatures. But at elevated temperatures, such as those occurring in oil refineries, petrochemical, and fertilizer plants, H_2S might initiate hydrogen embrittlement damage.

Yet another hydrogen-related problem is step-wise cracking, also called hydrogen blistering. This primarily occurs in low strength steels in particular pipeline steels (Refer Section 3.10.1). The main factors responsible for this type of failure are

1. Banded microstructures
2. High manganese levels in pearlitic bands
3. Manganese sulfide stringers
4. High sulfur level

Since this failure does not depend on residual and applied stresses, only the above factors need to be controlled. Normally the factors 3 and 4 are controlled to impart resistance of steel to this type of problem.

NACE standard RP0296-2004 provides Guidelines for Detection, Repair, and Mitigation of Cracking of Existing Petroleum Refinery Pressure Vessels in Wet H_2S Environments, while RP0472 deals with Methods and Controls to Prevent In-Service Environmental Cracking of Carbon Steel Weldments in Corrosive Petroleum Refining Environments.

NACE Standard MR0175 gives guideline for choosing carbon steels and corrosion resistant alloys for service against sulfide stress cracking by H_2S.

4.3.3 Caustic Stress Corrosion Cracking

This is a serious problem as alkaline (Caustic) solutions are frequently met with in chemical industries in various processing steps, typically the following:

- Neutralizing steps, where caustic (NaOH) solution is added to neutralize acidic pH to neutral/alkaline values
- Boiler/steam environments where the condensate is alkaline
- Waste heat boilers where pH of the feed water is kept alkaline by adding caustic
- Amine solutions such as aqueous monoethanolamine and diethanolamine solutions used for scrubbing acidic gases in petrochemical plants.

In all the above cases, caustic SCC occurs if the following two conditions are satisfied:

- Excessive evaporation occurs pushing the pH of the remaining solution to a highly alkaline value in the range 10–13 or if the caustic solution remains stagnant at certain pockets at the operating conditions and
- The steel portions in contact with such caustic environment is under tensile stress, either applied, as in pressure vessels, or residual like welds or near weld joints.

Such situations very often arise in steam condensate lines, shell side of vertical waste heat boilers, tube-to-tube sheet joints of heat exchangers where water is on the shell side.

The likelihood of SCC occurring under above conditions depends upon temperature and concentration of caustic solution contacting the surface. The limits of the above parameters for SCC susceptibility of carbon steel are shown earlier in Chapter 3 (refer Figure 3.19).

The crack mode in caustic SCC is mostly intergranular, but may become transgranular if contaminated with chlorides and/or cyanides.

If the service conditions, dimensions including thickness of the vessel and economics, demand only carbon steels for the above caustic application, then care should be taken to see that all the stressed portions of the vessel are thermally stress-relieved before putting the vessel into operation. The recommended temperature is 625 or 650°C. Thermal stress relief is the single most effective measure for preventing SCC in carbon steels in all the susceptible environments.

4.3.4   Favorable and Unfavorable Points in Using Carbon Steel as MOC

Points in favor of Carbon Steel as MOC

- Sufficient mechanical strength, ductility, and toughness for most of the applications
- Alloying technology, manufacturing/fabrication technology, welding technology, etc. and their effects on the final performance are well known and standardized
- Easy availability at the lowest cost both for initial installation and for replacement/maintenance
- Low but predictable corrosion resistance in most of the aqueous environments facilitating performance for a limited period using the concept of giving corrosion allowance during design
- Good thermal conductivity making carbon steels suitable for heat exchangers, condensers, and jacketed reactors where heat conduction across the wall is one of the service requirements.

Points not in favor of Carbon steel as MOC:

- Insufficient corrosion resistance in corrosive environments, necessitating protective linings, coatings, inhibitor additions, cathodic protection, etc. which add to the cost
- Susceptible to localized corrosion phenomena like SCC, Hydrogen Induced Cracking, etc. which calls for special chemistry change, heat treatment, hardness control, etc. which increases the cost still further keeping a finite possibility of the above phenomena occurring.

4.4   LOW ALLOY STEELS

The main driving force for the development of low alloy steels is to attain high strength and hardenability (the ease with which martensite structure is obtained without causing cracks). Over the years, several grades of low alloy steels have been developed to enhance high temperature creep resistance and temper embrittlement resistance. The important alloying elements are Cr, Mo, V, Nb, Ti, Ta, etc. The total alloy content of these steels are usually limited to 1–5 wt.%, though higher levels of Cr and Mo are also used to develop Cr–Mo steels. Low alloy steels are used in chemical process industry either for improved mechanical properties or for improved corrosion/oxidation/high temperature resistance. C–Mo and Cr–Mo steels are employed for their resistance to hydrogen-related corrosion phenomena at elevated temperatures; the choice depends upon H_2S partial pressure and temperature. Low alloy steels containing relatively high levels of chromium and molybdenum are used in high temperature applications where resistance to sulfur corrosion (sulfidation) is an additional requirement in addition to strength and resistance to hydrogen embrittlement. Table 4.3 lists the common relevant materials for high temperature use in chemical process industries, carbon and low alloy steels, and other materials to be discussed later in this chapter.

In applications where low alloy steels are considered, the corrosion characteristics of sulfur and hydrogen at elevated temperatures are of primary concern.

In the chemical process industry, the raw materials (the feed stocks) range from "sulfur free" intermediate products of the refineries such as ethylene, propylene, through "a few ppm of sulfur" light products such as natural gas and light distillates to "high sulfur," 5% or more, heavy fuel oils. Crude oil contains sulfur in the form of dissolved H_2S and mercaptans to a varying

TABLE 4.3   Common Materials of Construction for Elevated Temperature Use in Chemical Process Industries

Material	Scaling limit, °C*	Typical use
Carbon steel	540	General purpose
C-0.5Mo	540	Elevated temperature and hydrogen service
1Cr-0.5Mo	565	High pressure steam services and hydrogen services
1.25Cr-0.5Mo	565	
2Cr-1Mo	580	
5Cr-0.5Mo	620	For sulfur corrosion in liquid hydrocarbons or for hydrogen service
7Cr-0.5Mo	635	
9Cr-0.5Mo	650	
18Cr-8Ni austenitic stainless Steel	900	General corrosive service and high temperature sulfur and hydrogen
Incoloy 800 (18Cr-35Ni)	1100	High temperature and reformer furnaces
High carbon centrifugally cast 25Cr-20Ni	1100	

*Temperature above which scaling in ordinary flue gas, air, CO_2, or steam would be excessive, may also be limited by corrosion

degree. The oils containing less than about 0.5 wt-percent S are noncorrosive when heated and are known as "sweet crudes," and those containing sulfur more than the above level may be corrosive to carbon steel when heated and are termed "sour crudes." Addition of chromium to the steel would inhibit the breakdown of the sulfur compounds to hydrogen sulfide on the metallic surface and thereby reduce corrosion rates. Hence, the Cr containing alloy steels are extensively used in this application.

At this point a note of caution would be appropriate with regard to selection of low alloy steels for chemical process industries under circumstances which can promote hydrogen embrittlement. The tendency of this group of steels to hydrogen embrittlement is proportional to their strength. So their tolerance towards hydrogen content becomes weak as strength goes up. Figure 4.1 brings out this fact. Similarly, hardness has a great bearing on the susceptibility of steels to hydrogen embrittlement as will be discussed later. It is clear that the threshold level of hydrogen in steel and the extent to which low pH and high hydrogen sulfides can be tolerated in crude depend on these parameters.

It is necessary to point out some of the aspects related to fabrication, especially welding, of low alloy steels.

1. Fabrication such as stretching, bending, grinding, forging, and forming causes unwarranted residual tensile stresses.

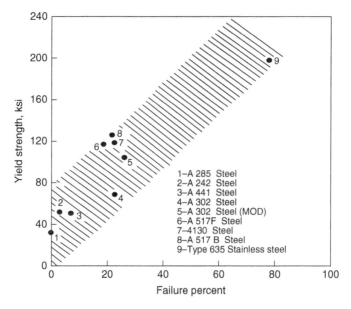

Figure 4.1   Relation of fracture frequency to yield strength (© NACE International 1969).

2. Welding can lead to microscopic variation in microstructures and mechanical properties that can hammer the useful properties of steel. Development of martensitic structures and rise in the hardness of the heat-affected zones (HAZ) and buildup of residual tensile stresses are causes of concern with respect to both SCC and hydrogen embrittlement.

It is very important to address these aspects adequately before the component is put into service. Stress relief or subcritical annealing is done in order to address the first aspect.

When steel is exposed to hydrogen, or hydrogen containing gases, at elevated pressures and at temperatures above about 250 °C, it may suffer the following two types of corrosion-related damages in addition to mechanical damages:

- Decarburization taking place at the external surface
- Embrittlement resulting in internal fissuring, lamination, and blistering leading to loss of strength and ductility, eventually unexpected fracture.

Both forms of damage are caused by the hydrogen reacting with the carbon of the steel to form methane gas. The latter cannot diffuse out; hence, coalesce to form microfissures and laminations (Read Section 3.10.1 for details).

The resistance of steel to decarburization is increased by the addition of carbide forming alloying elements such as Cr and Mo. The greater the amount of Cr, the higher the permissible operating limits of the steel. Such limits are found in the Nelson Chart given by API-941 specification.

Hydrogen-related damage is avoided by correct selection of materials in the design stage of the plant, using API 941.

Discussion related to weldments of steels, especially low alloy steels, is very appropriate. Both heat-affected zone as well as weld fusion zone are prone to cracking due to the following.

1. Hydrogen pickup
2. Martensite formation
3. Residual tensile stress

Welding electrodes are potential source of hydrogen. The moisture on the steel surface to be welded is another source for hydrogen. As pointed out earlier, the tolerance of steel to hydrogen embrittlement decreases as its strength and hardness levels go up. Grivelle (1976) related HE susceptibility of a weldment to HE between carbon content and carbon equivalent and it is a worthwhile reference to get an idea of the extent to which hydrogen affects weldments, based on the carbon equivalent of steels. According to the proposed diagram, as carbon content and CE equivalent increase (as represented by zone II and zone III in the diagram), extreme measures such as low hydrogen welding filler wire and pre- and post-weld heat treatments are needed to be taken to avoid HE or cold cracking of the weldments.

4.5 STAINLESS STEELS

Stainless steels are chromium containing iron base alloys, having superior corrosion resistance than steels. Chromium does the trick of imparting corrosion resistance through passive film formation. This was discovered independently by Harry Brearly of England and Eduard Meurer of Germany. Since then several developments have occurred on stainless steel family, giving rise to subfamilies. There are other elements such as molybdenum, nitrogen, silicon, and nickel which have found their place in stainless steels to supplement the corrosion-resistant property of chromium, more so with respect to localized corrosion, to a varying degree. But, the driving force for the development of new types of stainless steels came from not only corrosion resistance point of view but also other properties listed below.

- Corrosion/oxidation resistance.
- Mechanical properties.
- Fabrication (hot/cold working)
- Welding
- Low cost

Iron exists at room temperature in body center cubic structure (BCC), called ferritic phase. It exhibits a face-centered cubic phase (FCC), called austenitic phase, above 910 °C. Addition of nickel makes the iron to exhibit austenitic phase even at room temperature. The other elements, which are widely used in stainless steels, which impart a similar effect, are carbon, nitrogen, manganese, and copper. But, chromium and molybdenum (so also other less used elements such as Si) counter this effect by making the ferritic phase stable. So, by suitably alloying with these elements with right proportions, one can produce either a stainless steel with (a) ferritic phase or (b) austenitic phase or (c) a mixture of ferritic and austenitic phases. Thus, three families of stainless steels are developed based on the adjustment of alloy composition.

Yet another possibility, namely heat treatment, is used to produce two other families of stainless steels, viz., (d) martensitic and (e) precipitation hardenable (PH) stainless steels. Even for these types of stainless steels suitable alloying is important, though heat treatment is primarily responsible to obtain the desired phases. Martensitic and precipitation hardenable stainless steels offer attractive mechanical properties that are not possible with the three other types of stainless steels, but are obtained with a compromise on corrosion resistance. Between the two, martensitic and PH grades, the latter offers a better corrosion resistance with comparable or even better mechanical properties than the former. Some of the important properties of these families of stainless steels are briefly given below.

4.5.1 Ferritic/Martensitic Stainless Steels

Table 4.4 below gives a list of the most common ferritic/martensitic stainless steels which are in general used in chemical process industry. Ferritic stainless steels, like carbon steels and low alloy steels, are magnetic in character. Table 4.5 gives the UNS numbers and chemical compositions of these ferritic/martensitic Stainless Steels.

Type 410 SS is mainly employed as a lining material for pressure vessels handling sulfur bearing hydrocarbons at temperatures above 300 °C. It can

TABLE 4.4 Ferritic/Martensitic Stainless Steels

Material	Scaling Limit °C	Typical Uses
Type 410 SS, 12%Cr	705*	Lining of vessels and towers exposed to sulfur bearing hydrocarbons above 300 °C
Type 405 SS, 12%Cr-Al	815*	As above. Al is added to reduce hardenability
Type 430 SS, 17%Cr	815*	Heat exchanger tubes and tube sheets for corrosive service where SCC may be a problem with austenitic SS

*Stressed parts are limited to 400 °C because of temper embrittlement.

TABLE 4.5 Standard Ferritic/Martensitic Stainless Steels

AISI Type	UNS Designation	Composition, % Major Alloying Elements (Max or Otherwise Stated.)				
		C	Cr	Ni.	Mo.	Others
410 (Martensitic)	S 41000	0.15	11.50–13.50	–	–	–
405 (Ferritic)	S 40500	0.08	11.50–14.50	–	–	Al: 0.10–0.30
430 (Ferritic)	S 43000	0.12	16.00–18.00	–	–	–

also be considered for use as channels and channel covers in heat exchangers in the place of low alloy steel 5Cr–0.5Mo. Use of thick solid sections of 410 SS is not very common because the HAZ of fusion welds get embrittled to some degree and would be the place of crack initiation sites. Type 430 SS is mainly used for heat exchanger tubes with water on the shell side where corrosion resistance of lower Cr steels would be inadequate. Ferritic stainless steels are highly resistant to chloride SCC, hence considered for applications as cooler tubes handling water with appreciable amounts of chloride.

Ferritic stainless steels suffer from the following problems

1. Less versatile from welding stand point. They suffer from the formation of brittle intermetallic phases such as sigma.
2. Poor ductile brittle transition and so cannot be used for cryogenic applications.
3. Low ductility and formability
4. Poor resistance to hydrogen embrittlement

These problems are addressed in developing austenitic stainless steels. The properties of this family of stainless steels are discussed in the following section.

4.5.2   Austenitic Stainless Steels

This group of stainless steels, which are nonmagnetic, is the most important for process industry applications. Presence of considerable quantities of mainly nickel, sometimes with nitrogen and/or manganese, retains the high temperature austenitic phase at room temperature. This means that the austenitic stainless steels are more amenable for fabrication techniques like roll forming, drawing, welding, etc. at the same time retaining the general corrosion resistance corresponding to the level of chromium present.

Table 4.6 lists the most common austenitic stainless steels used for chemical process equipments and their nominal chemical compositions.

The grades mentioned in the above Table have a rare combination of corrosion resistance in most of the environments, high temperature strength and oxidation resistance, ductility and malleability, ease of fabrication, good weldability, adequate thermal conductivity making them suitable for heat exchanger tubes, and good impact resistance at low temperatures. The work-horse grade is Type 304 SS having a nominal chemical composition of 18 wt.% Cr, 8 wt.% Ni, and 0.08 wt.% max. C. All others are modifications on this to overcome certain specific limitations of the basic grade. Type

TABLE 4.6 Standard Austenitic Stainless Steels

AISI Type	UNS Designation	Composition, % Major Alloying Elements (Max. or Otherwise Stated)				
		C	Cr.	Ni.	Mo.	Others
304	S 30400	0.08	18.00–20.00	8.00–10.50	–	–
304L	S 30403	0.03	18.00–20.00	8.00–12.00	–	–
321	S 32100	0.08	17.00–19.00	9.00–12.00	–	Ti: $5 \times C$ min
347	S 34700	0.08	17.00–19.00	9.00–13.00	–	Nb: $10 \times C$ min
316	S 31600	0.08	16.00–18.00	10.00–14.00	2.00–3.00	
316L	S 31603	0.03	16.00–18.00	10.00–14.00	2.00–3.00	
316Ti	F 316Ti	0.08	16.50–18.50	10.50–13.50	2.00–2.50	Ti: $5 \times C$ min.to 0.70
317	S 31700	0.08	18–20	11–15	3.00–4.00	
317L	S 31703	0.03	18–20	11–15	3.00–4.00	
304H	S 30409	0.04–1.00	18–20	8–10.5	–	
310	S 31000	0.25	24–26	19–22	–	

89

304 SS has adequate corrosion resistance in almost all the environments. Its limitation is poor resistance to the following corrosion phenomena:

- general corrosion by reducing acids such as sulfuric, phosphoric, acetic acids,
- pitting, crevice corrosion, and SCC by chlorides and
- intergranular corrosion in the sensitized HAZ after welding.

Type 304L SS has carbon 0.03 wt.% max. instead of 0.08 wt.% max, making it better resistant to heat-affected zone intergranular corrosion. Unfortunately reducing the carbon level reduces the mechanical strength also. To overcome this, the grades Type 321 and Type 347 were developed. These are known as stabilized grades and are having alloying elements Ti and Nb (Cb), respectively, which have higher carbide-forming tendencies in preference to that of chromium, thereby leaving the latter free for corrosion resistance purpose. These have good resistance to intergranular corrosion in the weld HAZ, without losing mechanical strength. It is necessary to point out the fact that these stabilized grade stainless steels are prone to knifeline attack—a form of intergranular corrosion—that occurs very close to weld-fusion zone. Knifeline attack occurs on those cases where multipass welding is carried out or the welded structure is subjected to the temperatures in the sensitization range.

Type 316 SS has Mo as an extra alloying element at the level of 2–3 wt.%. This makes the steel more resistant to general corrosion by reducing acids and pitting and crevice corrosion by chlorides. Because Mo at the level of 2–3% is a major substitutional alloying element, Type 316 SS is mechanically marginally stronger than Type 304 SS. Type 316L SS is the low carbon variety of Type 316 making it resistant to intergranular corrosion also. Type 316Ti is the stabilized variety over 316 SS, thereby making it resistant to intergranular corrosion and at the same time without losing the basic strength of 316 SS. Types 317 and 317L SS are improvements over 316 and 316L SS with respect to the level of Mo, 3–4% instead of 2–3%, thereby further enhancing the resistance to corrosion by reducing acids and chlorides. Addition of nitrogen leads to a rise in resistance to pitting, crevice, and intergranular corrosion. In fact an empirical relation has been established to relate pitting resistance (in terms of pitting-resistant equivalent number, PREN) with the chemical composition.

$$PREN = \%Cr + 3\%Mo + 16\%N$$

The efficiency of these elements in offering resistance to pitting, however, decreases if the stainless steel is welded. Welding leads to microsegregation of alloying elements which is responsible for lowering the pitting resistance

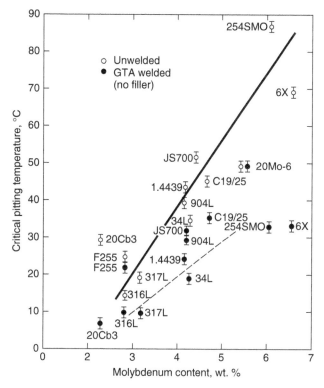

Figure 4.2 Variation of the critical pitting temperature (determined from ferric chloride solution) with the molybdenum content of various stainless steels. (Source: Garner 1985, Welding Journal, Vol 62, No. 1.)

of weld fusion zone. Figure 4.2 shows that higher the alloying elements, larger are the detrimental effect of welding.

To summarize the corrosion-resistant grades, 304 SS is still the popular grade for overall corrosion resistance in moderately corrosive environments. The next popular grade in somewhat more corrosive environments is 316L SS. But all the above grades are susceptible to the phenomenon of SCC by chloride, which is avoidable by using "Duplex Stainless Steel" to be discussed later in this chapter.

Type 304 has somewhat inadequate mechanical properties at temperatures above 500°C and at still higher temperatures, though retaining the corrosion/oxidation resistance. To overcome these limitations, the grades 304H, 316H, and 310 were developed. The H grades have a minimum carbon level at 0.04 wt.% and maximum carbon level at 1.0 wt.%, whereas 304 and 316 have no minimum carbon specified and have maximum carbon level at 0.08 wt.%. The 310 grade has very high chromium level, at around 20 wt.%, and very high Ni content, at around 25 wt.%, making it highly resistant to very high temperature corrosion phenomena such as oxidation, sulfidation, carburiza-tion, etc., thereby allowing them to be used for high temperature equipments like reformers and furnaces.

The grades of austenitic SS discussed so far are in the *wrought* form, mean-ing that the products have been deformed, rolled, or deformed, after casting and annealing. On the other hand, *cast* part is not rolled or deformed after cast-ing into shape. Cast stainless steels have somewhat different chemical com-position than their wrought counterparts, making them suitable to be retained in the cast form and at the same time maintaining the equivalent corrosion resistance and other properties such as high temperature strength. One major difference is higher silicon content which increases castability which is nec-essary for casting thin sections and small parts. Another difference is in the level of ferrite formers such as chromium and molybdenum, which substan-tially increase the strength with corresponding increase in austenite retainers such as Ni and C. These modifications impart high temperature strength and oxidation resistance. Typical cast grade applied in chemical process industry is HK 40, the cast equivalent of 310SS, used for high temperature reformer tubes in oil refineries, petrochemicals, and fertilizers.

4.5.3   Super Austenitic Stainless Steels

Super Austenitic Stainless Steels, also called High Alloy Austenitic Stain-less Steels, and certain nickel base alloys are developed for much greater corrosion resistance than the 300 series austenitic stainless steels, particu-larly against chloride pitting in solutions of high chloride with or without some acidity, including sea water. They also possess, with or without further modifications, improved resistance to reducing acids, like sulfuric, phospho-ric, and acetic acids. These developments were through increasing Cr, Ni, and Mo contents, adding intentionally high quantities of dissolved nitrogen thereby maintaining the austenitic structure and keeping the carbon content fairly low, thereby maintaining the resistance to intergranular corrosion. Some have copper added for increased corrosion resistance to sulfuric and phospho-ric acids. Similarly titanium and niobium are also added as stabilizing agents against heat-affected zone intergranular corrosion. Table 4.7 lists some of the common super austenitic stainless steels and their nominal compositions. It is generally noted that austenitic, ferritic, and duplex stainless steels (DSS) qualify themselves to be "super"-austenitic, ferritic, and duplex when their PREN number given by the above equation exceeds 40.

316LN SS was developed specifically to withstand high temperature and high pressure carbamate solution in urea production, in the stripping section where the unreacted CO_2 is stripped from the product urea solution. 904L is considered the work-horse of the super austenitic stainless steels having

TABLE 4.7 Super Austenitic Stainless Steels

Common Name	UNS Designation	Composition, wt. % Major Alloying Elements (Max. or Otherwise Stated)				
		C	Cr.	Ni.	Mo.	Others
904 L	N08904	0.02	19–23	23–28	4.00–5.00	Cu: 1.00–2.00
Alloy 28	N08028	0.03	26–28	29.50–32.50	3.00–4.00	Cu: 0.60–1.40
Alloy-6XN	N08367	0.03	20–22	23.50–25.50	6.00–7.00	N: 0.18–0.25
316LN	S31653	0.03	16–18	10–14	2.00–3.00	N: 0.10–0.16
20Cb – 3	N08020	0.07	19–21	32–38	2.00–3.00	Cu: 3.00–4.00, Nb: $8 \times$ C
254SMO	S31254	0.02	19.50–20.50	17.50–18.50	6.00–6.50	Cu: 0.50–1.00, N: 0.18–0.22

satisfactorily good resistance to many of the acids and chloride solutions. Alloy 28 has been developed to withstand phosphoric acid corrosion in phosphoric acid manufacturing plants, both as heat exchanger tubes and as reactor parts. 20Cb-3 is one of the earlier, first-generation super austenitics, developed for resistance to sulfuric acid by increasing the nickel content to a high level and by adding copper as an extra alloying element. Alloy 6XN and 254SMO are commonly known as 6Mo grades and have been developed to withstand pitting and crevice corrosion resistance by high concentration chloride solutions.

4.5.4   Duplex Stainless Steels

Almost all the standard austenitic stainless steels suffer from the deficiency of poor resistance to SCC induced by several anions, particularly chlorides. Super austenitic stainless steels are resistant to SCC, but they are quite expensive since they have been developed by increasing the content of nickel which is the most expensive, and cost deciding, alloying element. In order to overcome this limitation, DSS have been developed.

Published reports (Copson 1959, Speidel 1981) show that SCC resistance is at a minimum at the nickel content of 8 to 10%, that corresponding to Type 304 SS. The resistance increases when nickel becomes either lower or higher than the above range. The cost advantage of lower nickel has been used in developing the DSS which have much lower Ni than standard austenitics with somewhat higher Mo and N. These modifications result in a duplex microstructure of partly austenite and ferrite. The latter phases form alternate bands. Since, as mentioned under ferritic stainless steels, the ferrite phase is resistant to SCC, the duplex structure shows full resistance to SCC in spite of the presence of alternate bands of austenite. High Mo gives higher resistance to pitting and crevice corrosion.

High nitrogen (austenite stabilizing alloying element cheaper than nickel) helps in retaining the austenite phase, thereby making the steel more amenable for fabrication as compared to ferritic stainless steels. The latter alloying elements also impart high mechanical strength and toughness without losing much of ductility. DSS have yield strength almost double that of the standard austenitics making the usage of thinner sections possible, reducing the cost still further. Table 4.8 gives compositions of typical DSS. Type 329 and 3RE60 are the first generation duplex grade, developed with the aim of achieving the austenite ferrite phase balance, approximately 50/50. 2304 and 2205 are developments on the above, giving rise to improved phase stabilities particularly at the welds and HAZ and also improved corrosion resistance at more and more higher temperatures and chloride concentrations. Presently, 2205 is the most popularly used grade. The last three listed in the

TABLE 4.8 Duplex Stainless Steels

Common Name	UNS Designation	Composition, wt % Major Alloying Elements (Max. or Otherwise Stated)				
		C	Cr.	Ni.	Mo.	Others
AISI-329		0.10	25.00 – 30.00	3.00 – 6.00	1.00 – 2.00	
3RE60	S31500	0.03	18.00 – 19.00	4.25 – 5.25	2.50 – 3.00	
2304	S32304	0.03	21.5–24.5	3.0–5.55	0.05–0.06	
2205	S31803	0.03	21–23	4.5–6.5	2.5–3.5	N: 0.08–0.20
2507	S32750	0.03	25.00	7.00	4.00	N: 0.30
Ferralium 255	S32550	0.04	24–27	4.5–6.5	2.00–4.00	Cu: 1.50–2.50, N: 0.10–0.25
Zeron 100	S32760	0.03	25.00	7.00	3.00	N: 0.30
3207	S33207	0.03	27.00	6.5	4.8	N: 0.4, Co=1.0

Table are "super duplexes" having exceptionally high resistance to sea water corrosion phenomena including general and pitting corrosion and SCC. Field experience shows that 2507 is, however, prone to crevice conditions when used in heat exchangers, coolers, and condensers, employing sea water as a coolant. To circumvent this problem, hyper DSS are being commercialized. Duplex stainless steels cost less than either nickel base alloys or super austenitic stainless if one were to use an equivalent PREN in all these types of alloys.

4.6 NICKEL BASE ALLOYS

Nickel and its alloys offer a wide range of corrosion resistance. Nickel can accommodate larger amounts of alloying elements, particularly chromium, molybdenum, copper, and also iron, in solid solution than iron. Hence, nickel base alloys are much more corrosion-resistant than stainless steels and can be used in environments more severe than those for which stainless steels including the super austenitics and the duplexes are specified. This advantage is carried forward to higher temperatures also; therefore nickel alloys are preferred over stainless steels for high temperature corrosive applications. The latter include process gases and furnace atmospheres, with and without condensation containing acidic phases. Further, these alloys also have alloying elements such as aluminum, titanium, and niobium which stabilize precipitates and strengthen high temperature properties including fabrication and welding characteristics. Table 4.9 gives typical chemical compositions of certain popular nickel base alloys in use in chemical process industry. All the alloys in the above table are highly corrosion-resistant in all the environments including sea water, acidic chloride, caustics, etc. The alloys Incoloy 800/800H and Inconel 625 are also resistant to high temperature gases and corrosives. Monel 400 is particularly used as cladding on carbon steel evaporators in the production of fresh water from sea water.

4.7 COPPER BASE ALLOYS

Copper and copper base alloys find applications as process equipment materials of construction in chemical process industries basically for the following reasons:

- their overall good corrosion resistance in general due to their "noble" position in the EMF series
- their superior thermal (and hence electrical) conductivity values

TABLE 4.9   Corrosion & Heat-Resistant Nickel Base Alloys

Common Name	UNS Designation	Composition: wt. % Major Elements (max. or otherwise stated)							
		C	Cr	Cu	Fe	Mo	Ni	Ti	Others
Monel Alloy 400	N04400	0.15		31.5	1.25		Bal.		Si: 0.50
Ally B – 2	N10665	0.01	1.0		2.00	28.0	Bal		Si: 0.10
Inconel 600	N06600	0.08	16.0	0.50	8.00		Bal	0.30	Si: 0.50
Inconel 601	N06601		23.0		14.1		Bal		Al: 1.35
Incoloy 800	N08800	0.1	21.0	0.75	44.0		32.5	0.38	Si: 1.00
Incoloy 800H	N08810	0.08	21.0	0.75	44.0		32.5	0.38	Si: 1.00
Incoloy 825	N08825	0.05	21.5	2.0	29.0	3.0	42.0	1.00	Si: 0.50
Inconel 625	N06625	0.10	21.5		5.00	9.00	62.0		Nb: 4.00 Si: 0.50
Alloy C-276	N10276	0.01	15.5		5.5	16.0	57.0		Si: 0.08 W: 4.00
Alloy C – 22	N06022	0.015	22.0		3.0	13.0	56		Si: 0.08 W: 3.00

- their easy amenability for all fabrication techniques and
- their resistance to biofouling by microorganisms.

Hence, they provide superior service for the following industrial applications:

- Chemical process equipments involving highly corrosive, but deaerated, acids, and other inorganic and organic solutions
- Freshwater and sea water applications such as piping and pipe components and plumbing fittings
- Heat exchanger and condenser tubes and tube-sheets in steam power plants, oil refineries exchanging heat between hydrocarbon streams.

In addition to copper metal itself being used as the MOC, quite a few of its alloys find applications in CPI. A few of these alloy groups, the important alloying element in the respective group, and their most common applications are the following:

- Pure copper as tubing for steam tracing and other somewhat corrosive applications not demanding high mechanical strength

- Brasses containing Zn at various levels for heat exchanger tubes, particularly in sugar plants for cane juice evaporators, with small quantity of Al as "aluminum brass" for sea water cooler/condenser tubes in coastal power stations and with small quantity of Sn as "admiralty metal" tubes for fresh water-cooled condensers in inland power stations
- Bronzes containing Sn, Al, and P in various levels for certain plates and forgings such as tube sheets in heat exchangers requiring high thickness and galvanic compatibility with tubes of copper base alloys. Aluminum bronze, in specific, containing aluminum only as the alloying element in the range 5 to 12%, has good erosion and impingement resistance and good corrosion resistance to sulfite solutions
- Cupro-nickels containing Ni as the main alloying element (10 or 30%) with or without Fe for condenser and cooler tubes in sea water applications requiring more corrosion resistance and higher mechanical strength and more resistance to erosion and impingement than Aluminum brass tubes.

As far as corrosion characteristics are concerned, copper and copper base alloys are not completely free from certain forms of corrosion. The corrosion aspects are briefly discussed below:

Uniform corrosion, resulting in thinning at unacceptable rates, might occur in oxidizing acids, sulfur-bearing compounds, and solutions like polluted sea water, ammoniacal atmospheres, and aqueous solutions containing ammonia and also in cyanide solutions.

Galvanic corrosion does not occur in copper base alloys since they invariably form the cathodic component of the couple. But the galvanic contact with copper base alloys would accelerate the corrosion rate of the anodic member of the couple. This needs to be properly designed out.

Impingement and erosion corrosion, the so-called inlet end corrosion, occur in condensers and coolers where rapidly moving turbulent water breaks the protective film locally and leads to erosive attack. Inlet-end erosion is usually reduced or overcome with the help of inlet ferrules made of plastic and/or soft compatible metal which are inserted at the tube inlets to the extent possible without affecting heat transfer across the wall of the tubes.

Dealloying, like dezincification, discussed earlier in Chapter 3, does occur in copper base alloys with tin, aluminum, and nickel also. It frequently occurs in waters high in chloride, carbon dioxide, and acidity and also with stagnant waters. Inhibitor alloying additions such as arsenic, antimony, or phosphorus to the base alloy such as aluminum brass or admiralty metal are commonly made for increasing the resistance to dezincification.

Stress Corrosion Cracking (SCC), also discussed at length in Chapter 3, occurs in copper base alloys also, though not to the same extent as that in

TABLE 4.10 Copper and Typical Copper Base Alloys

Sr. No.	Generic Name	UNS Number	Main Elements
1	Coppers	C10100-C15815	>99% Cu
2	Brasses	C21000-C28000	Cu–Zn
3	Tin Brasses	C40400-C48600	Cu–Zn–Sn–Pb
4	Phosphor Bronzes	C50100-C52400	Cu–Sn–P
5	Aluminum Bronzes	C60800-C64210	Cu–Al–Ni–Fe–Si–Sn
6	Copper-nickels	C70100-C72420	Cu–Ni–Fe

austenitic stainless steels. Ammonia and ammonium compounds are the common corrosive agents most often associated with SCC of copper base alloys, in the presence of moisture, oxygen, and carbon dioxide. SCC mostly occurs near the tube ends where residual stresses are present due to roll expansion of the tubes into the tube-sheets. Minimizing the residual stresses is the single most controlling step against SCC in copper base alloys.

Biofouling in waters, particularly in sea water, a common phenomenon leading to localized corrosion pitting resulting in leakage even in stainless steels and nickel base alloys, does not easily occur in copper base alloys. This advantageous property is probably due to some poisonous characteristics of the leached heavy metal copper ions (cupric) present in the water. In this respect copper alloys are even better than titanium which otherwise, as shown under the section on titanium, is a superior corrosion-resistant MOC for chemical process industries.

The common materials of construction within the copper and copper base alloys group, which are encountered in CPI, are shown in Table 4.10.

4.8 TITANIUM

Titanium is light, strong, and corrosion-resistant. Its specific gravity, 4.5, is about 60% that of steel and stainless steel and only half that of copper and nickel. Thus, titanium becomes very handy for "weight watchers" among the designers, manufacturers, and users of process equipments. The mechanical strength of pure titanium is greater than that of carbon steels and stainless steels. This allows the designers to scale down the thickness of the equipments further by resulting in lower equipment cost.

The corrosion resistance of titanium is superior to stainless steels in many ways; particularly its resistance to chloride solutions is as "good" as that of platinum. It is also highly corrosion-resistant to all aqueous environments with varying degrees of oxidizing power and in gaseous environments with small amounts of moisture and oxygen. By the same reasoning, anhydrous conditions in the absence of a source of moisture/oxygen may result in severe corrosion of titanium. The well known incompatibility of titanium to dry chlorine is a common example of such a behavior. Another area of concern with respect to titanium is its poor corrosion resistance to crevice corrosion in low pH high sodium chloride concentration brines at temperatures above about 80 °C.

Titanium finds extensive usage in nitrogenous fertilizer industries for handling the various process streams in urea plants including strip lining in urea reactors, heat exchanger tubes in urea strippers, and carbamate condensers. Titanium is also used as process equipments for bleaching of pulp in pulp and paper plants, involving highly bleaching agents like HCl, ClO_2, etc. Titanium finds application also in soda ash industries as carbonation tower coolers, cooling carbonated ammoniacal brine solution on the shell side with sea water on the tube side. Another major use of titanium is in the form of very thin corrugated sheets used as elements in Plate Type Heat Exchangers/ Coolers (PHE) used for cooling purposes in chemical process industries. ASTM B338-14 covers Standard Specification for Seamless and Welded Titanium and titanium alloy tubes for Condensers and Heat Exchangers. Some of the prominent among them are: Grade 1 (UNS R50250) unalloyed titanium, Grade 2 (UNS R50400) unalloyed titanium, and Grade 7 (UNS R52400) unalloyed titanium plus 0.12 to 0.25% palladium. Variation in Pd content with different grades is also developed.

Perhaps one of the major disadvantage of using titanium and its alloys can arise from the fact that they are vulnerable to hydrogen embrittlement. So, titanium structures are never cathodically protected and should not be a part of any structure that is cathodically protected.

4.9   ALUMINUM ALLOYS

Aluminum exhibits high thermodynamical driving force for corrosion as reflected by its very active standard potential ($E° = -1.7\,V$). However, it forms passive oxide film on exposure to general atmosphere and aqueous solution. As a result, aluminum and its alloys show very high resistance to uniform corrosion, as well as to atmospheric corrosion. Aluminum and its alloys are, however, prone to localized corrosion, such as pitting and crevice, even if low levels of chlorides are present in an environment. The localized corrosion will be further accelerated by the presence of relatively noble metallic ions such as Cu^{2+}, Fe^{3+}, Hg^{2+} in the process medium, as they bring down the pitting potential. Aluminum and its alloys are strongly attacked by acidic as well as alkaline environments and so cannot be used under these corrosive conditions. Furthermore, aluminum and its alloys could become anodic to several other metals and alloys should they be joined with them in any structural component and thereby suffer galvanic corrosion. For these reasons aluminum and its alloys do not find as much applications in process industry as they do in transportation industries.

The major advantage of using aluminum and its alloys in preference to commonly used steels and stainless steels is their high specific strength (=strength/density), and in some cases, their high electrical and thermal conductivity. In addition, it is nonmagnetic, nontoxic, and can be used for cryogenic applications without losing its strength. In order to increase specific strength of aluminum, it is alloyed with various metallic elements such as Cu, Mn, Mg, Zn, Si, and Li and thus various types of aluminum alloys have been evolved (Table 4.11).

Somewhat similar to stainless steels, aluminum alloys are designated based on the composition and there are eight of them. But what differentiates stainless steels (except that of precipitation-hardened (PH) stainless steels) from that of aluminum alloys is that the strength of the latter can be significantly enhanced through heat treatment process and that of the former is not. This leads to additional designation to the alloys to indicate the nature of treatments. Typically, aluminum alloys are heat-treated (called aging treatment) to what are called as under aged, peak aged, and overaged treatments. The book on aluminum and aluminum alloys edited by Davis (1993) gives a good overview of various aspects of aluminum alloys. While the peak aged treatment (T6) provides the highest strength, it also makes the alloy more susceptible to stress corrosion. For these reasons (mainly due to SCC), the alloys

TABLE 4.11   Classification of Wrought Aluminum Alloys and Their Mechanical Strength Range

Designation	Alloying Elements	Strength Range, MPa
1XXX	Aluminum, ≥99.00%	70–175 (c)
2XXX	Copper	170–520 (ht) (depends on Cu an Si levels)
3XXX	Manganese	140–280 (c)
4XXX	Silicon	105–350 (c)
5XXX	Magnesium	140–280 (c)
6XXX	Magnesium and silicon	150–380 (ht)
7XXX	Zinc	380–620 (ht) (depends on Cu)
8XXX	Other elements	280–560 (ht) (Li alloyed)

c = cold-worked and ht = heat-treated conditions

are overaged (to tempers such as T7) to optimize the alloy strength as well as its resistance to SCC. This optimization has led to development of other treatments such as multistep aging (Bobby et al. (2003)) as well as retro-gressive re-aging treatment. Another method of improving SCC resistance without losing much of the strength of aluminum alloy is by lowering recrys-tallization (Bobby et al. (2005)). At this point it is necessary to point out the fact that the added alloying elements form various types of precipitates ($CuAl_2$, $CuMgAl_2$, Mg_2Si, Al_8Mg_5, Al_6 (Mn,Fe), $Al_{12}Mg_2Cr$, Al–Li) that are required to enhance the alloy strength. Unfortunately, barring a few (such as Al_6(Mn,Fe), $Al_{12}Mg_2Cr$, Mg_2Si) these precipitates behave either anodic or cathodic to the aluminum matrix (of the alloy) leading to loss in pitting corrosion, exfoliation corrosion, and stress corrosion resistance. But they are the necessary evils of the alloy as they promote their strength.

There are several ways by which the corrosion resistance of aluminum alloys can be increased (Wernick et al. 1987). Corrosion being a surface process, most of these ways are directed towards surface treatment. Any of the processes, namely (a) cladding with pure aluminum or Al–Mn thin layer, (b) anodizing, (c) plasma electrolytic oxidation, (e) treating with chromates, molybdates, cobaldates, and permanganates, (f) protective organic coatings, (g) porcelain enameling (h) electroplating, and (i) electroless plating of metal, can be employed depending on the applications to prevent corrosion of aluminum alloys. Mn addition through alloying and heat treatments is also employed to control corrosion.

Among the various classes of aluminum alloys, 5XXX series seem to exhibit the highest resistance to corrosion, while 2XXX and 7XXX alloys need external protection to keep their corrosion tendency within some acceptable limit. The corrosion resistance behavior of 1XXX, 3XXX, and 6XXX alloys seems to exhibit moderate corrosion resistance. A more detailed treatment of this subject can be found in an exclusive book "Corrosion of aluminum and alloys" edited by Davis (1999).

4.10   NONMETALLIC MATERIALS

Ceramics/Inorganic oxide glasses and polymers are other materials used for structural applications. Compared to metals, they exhibit better corrosion resistance. However, they find limited application as MOC. This is primarily due to the fact that most of the ceramics have poor toughness and thermal conductivity and the polymers suffer high temperature structural instability. The ceramics and polymers find extensive applications as protective coatings. They do suffer from chemical degradation. Hence, this aspect is briefly covered here.

4.11   CERAMICS/INORGANIC OXIDE GLASSES

Ceramics are brittle and do not resist impact loading and so find limited applications for structures. Richlen and Parks (1991) listed some of the ceramics
used for corrosion-resistant heat exchangers which operate at high temperatures. However, most of the ceramics, for the purpose of enhancing corrosion
protection, are used as refractories and coatings. Yttria-stabilized zirconia
ceramic is widely used as thermal barrier coatings for gas engines. Yttria,
used to stabilize the tetragonal phase, has been found to leach out during
high temperature corrosion such as hot corrosion (Sreedhar and Raja (2010)).
To improve resistance to high temperature corrosion, several modifications
to zirconia coating are carried out. In contrast to oxide ceramics such as
zirconia, nitrides and carbides are thermodynamically unstable in high temperature oxygen-containing environments. Nevertheless, these ceramics can
form protective oxides such as SiO_2. The book on Corrosion of Ceramic and
Composite Materials by McCauley (1994) provides a detailed account of the
corrosion behavior of these materials.

Porcelain enamels are other inorganic oxide materials that find extensive
applications as coatings applied over fabricated parts of metals such as steel,
cast iron, aluminum, etc. These materials are glassy in nature and are used
for ambient and elevated temperature ($110–165\,°C$) applications. Porcelain
enamels are highly resistant to a variety of chemicals including alkaline
solutions, all acids except hydrofluoric acid, concentrated sulfuric, nitric and
hydrochloric acids.

Ceramic lining (chemical-setting) can be applied on any complicated
installations. They are resistant to high- and low-treated concentrated acids
(H_2SO_4, HNO_3, HCl, H_3PO_4). Potassium silicate (K_2SiO_3) cements are now
widely used for various applications such as scrubbers, ducts, chimneys,
tanks, reactions vessels, floors, and trenches in several industries.

4.12   ORGANIC POLYMERS/PLASTICS

Polymers offer attractive properties such as excellent corrosion resistance,
light weight, ease of fabrication, and installation that make them good candi-
date materials for a variety of chemical process applications. Structures made
of polymers require less maintenance than those made of metals. Though
numerous polymers are synthesized, only a few of them are useful for structural applications that are made into plastics (obtained with some additives).
Plastics can be classified into two major categories, namely thermosets and
thermoplastics. The former type undergoes chemical reaction on heating and
cannot regain their original state once the reaction is completed. Thermoplas-
tic, on the other hand, can be remelted and reused. Between thermosets and

thermoplastics, the former is more resistant to chemical degradation than the latter. Thermoplastics can be further classified as amorphous and crystalline materials. Generally, between them, crystalline thermoplastics are more resistant to corrosion than their amorphous counterparts. They are more susceptible to wear, stress cracking, and fatigue cracking than the latter. Another problem of concern in using plastics for structural applications is their mechanical integrity with temperature. Handbooks by De Renzo (1985), Rotheiser (2000), and Wilks (2001) provide extensive information of the properties and structural application of polymers. Chawala and Gupta (1993) provide a summary of relative corrosion resistance of normal polymers and elastomers, in most widely used environments (Table 4.12).

Polymer structure, which renders various properties, is degraded due to hydrolysis, oxidation, thermal, mechanical, radiation, and chemical and biological attack. Notably, environments can also cause SCC. Polyethylene, polystyrene, and ABS are prone to such failures. In general, larger the saturation larger is the stability of polymers and larger the functional groups lesser will be the stability. Polymers are alloyed to gain synergy in the property, while each of these polymers may be better in some properties, may lack in another. Various additives such as stabilizers, anti-oxidants, anti-aging, fillers, reinforcements, and coupling agents can also be selectively added to enhance specific property of the polymers. Thermoplastics used for corrosion resistance are fluorocarbons, acrylics, nylon, chlorinated polyether, polyethylene, polypropylene, polystyrene, and polyvinyl (chloride). As far as the thermosets are concerned, epoxies, phenolics, polyesters, silicones, and ureas are resistant to corrosion.

A special mention of fluorocarbons needs to be made as they offer not only outstanding chemical resistance, but also exhibit good elastic properties over a range of temperatures. So they are used for a variety of applications as sealer, linings, tubing, hose, belting, fabrics, caulks, adhesives, dampeners, etc.

Plastics soften at high temperatures and swell on long-term exposure to environment. From this point of view, metals are quite superior. One of the ways to utilize excellent corrosion resistance of plastics is to use them as protective coatings, linings for metallic structures. The book by Munger (1984) provides an extensive coverage of this topic.

4.13   MATERIALS SELECTION FOR CORROSION PREVENTION IN HYDROCARBON SERVICE

The authors had an opportunity to study the overall materials selection requirements unique to hydrocarbon processing equipments in connection with a specific request of a corrosion-related study. A summary of the above

TABLE 4.12 Rating of Chemical Corrosion Resistance of Polymers and Elastomers, Based on an Arbitrary Scale of Increasing Resistance of 0–10 (Chawala and Gupta (1993))

Polymers	Acids	Alkali	Oxidizing agents	Oils, greases, petroleum, and fuels	Hydrocarbons	Average value
PTFE, FEP, PFA(b)	10	10	10	10	10	10
Kalrez fluoroelastomer	10	10	10	10	10	10
ETFE. ECTFE(c)	10	10	10	10	8	9.6
PVDF(d)	10	10	10	10	7	9.4
Other fluoroelastomers	9	7	10	9	8	8.8
PCTFE, PVF(e)	10	10	10	8	5	8.6
PVC-unplasticized PVC, PVDC-PVC copolymer, CPVC(f)	9	10	10	9	5	8.6
Furan resins	10	10	1	10	10	8.2
Poly(phenylene sulfide)	6	8	4	10	10	7.6
Thermosetting polyimides	6	6	8	8	9	7.4
Poly(ether sulfone), polyarylketone	8	9	6	8	6	7.4
Epoxy resins	7	8	3	9	9	7.4
Polysulfone	8	8	7	6	6	7.0
Nitrile rubber	7	8	4	10	5	6.8
Acetal resins	5	10	4	8	9	6.8
Chloroprene polymers	9	8	4	8	5	6.8
Unsaturated polyester, FRP(g)	7	5	5	8	8	6.6

(continued)

TABLE 4.12 (*Continued*)

Polymers	Acids	Alkali	Oxidizing agents	Oils, greases, petroleum, and fuels	Hydrocarbons	Average value
Polyolefms(h)	8	9	7	6	3	6.6
Polyacrylic rubber	6	6	6	9	6	6.6
Chlorosulfonated rubber	10	7	7	5	4	6.6
Polysulfide rubber	4	8	4	7	9	6.4
Silicone rubber	7	9	7	5	2	6.2
Poly(amide-imide), Polyetherimide, Thermoplastic polyimides	8	5	2	8	8	6.2
Polypropylene, Ethylene-propylene elastomers	9	9	6	5	2	6.2
Thermoplastic polyesters	6	5	4	8	8	6.2
Butyl rubber	10	10	4	4	2	6.0
Nylon	3	8	3	8	8	6.0

Abbreviations: FEP = Fluorinated ethylene propylene copolymer; ETFE = Ethylene-tetrafluoroethylene copolymer; ECTFE = Ethylene-chlorotrifluoroethylene copolymer; FRP = Fiber-reinforced plastics except polypropylene; PCTFE = Polychlorotrifluoroethylene; PFA = perfluoroalkoxy resins; PTFE = Polytetrafluoroethylene; PVDF = Poly(vinylidene fluoride); PVF = Poly vinyl fluoride; PVC = Poly(vinyl chloride); PVDC = Poly(vinylidene chloride); CPVC = Chlorinated poly(vinyl chloride) (Reprinted with permission from ASM International. All rights reserved www.asminternational.org)

overall material requirement study unique to hydrocarbon industries is given in this section.

Materials selection for corrosion prevention in Hydrocarbon service, gas, oil, or any downstream hydrocarbon essentially depends upon the content of moisture/water, sulfur (sweet or sour), dissolved carbon dioxide, and dissolved chloride. Hydrocarbons themselves are not corrosive. Materials selection for sweet hydrocarbons is fairly straight forward based on conventional methods to choose against the above mentioned inorganic corrosives. On the other hand, materials selection for sour service is somewhat complicated and is basically guided by NACE Standard MR0175: "Standard Material Requirements—Sulfide Stress Cracking Resistant Metallic Materials for Oilfield Equipment".

In this section, a brief summary of the Materials Selection Guidelines as per the above Standard is given.

4.13.1   Materials Selection as per NACE MR0175

4.13.1.1   Carbon and Low Alloy Steels   All carbon and low alloy steels are acceptable at 22HRC maximum hardness provided they contain less than 1% nickel and are used either in the hot-rolled or cold-worked and heat-treated condition. The exception to this general acceptance is the use of free-machining steels to which intentional sulfur/selenium is added.

4.13.1.2   Cast Iron   Gray, austenitic, and white cast irons are not acceptable for use as pressure-containing member. These materials may be used in internal components.

4.13.1.3   Austenitic Stainless Steels   Austenitic stainless steels, those listed in the Standard, either cast or wrought, are acceptable at a hardness of 22 HRC maximum in the annealed condition, provided they are free of cold work designed to enhance their mechanical properties. Certain high alloy super austenitic stainless steels (UNS S31254) are acceptable in the annealed or cold-worked condition at a hardness level of 35HRC maximum.

4.13.1.4   Ferritic and Martensitic Stainless Steels   These are acceptable at a 22HRC maximum hardness, provided they are in annealed/heat-treated condition as per the Standard.

4.13.1.5   Duplex Stainless Steels   The wrought duplex (austenitic/ferritic) stainless steels listed in the Standard are acceptable at 28HRC maximum in the solution annealed condition. The most common duplex, UNS S31803 (2205), is acceptable in the solution annealed and cold-worked condition if the

TABLE 4.13   Hardness Limits for Non-ferrous Alloys for Hydrocarbon Service

Alloy	Hardness, Maximum HRC
Monel 400 (Ni–Cu alloy)	35
Monel K–500(Ni–Cu alloy)	35
Incoloy 800 (Ni–Fe–Cr alloy)	35
Incoloy 825 (Ni–Fe–Cr–Mo alloy)	35
Alloy-28 (Ni–Fe–Cr–Mo alloy)	33
Alloy 30 (Ni–Fe–Cr–Mo–W alloy)	41
Inconel 600 (Ni–Cr alloy)	35
Inconel 625 (Ni–Cr–Mo alloy)	35
Hastelloy C-276(Ni–Cr–Mo alloy)	45
Cast nickel base alloys	22 (in general)
Pure tantalum	55
Titanium alloys	See the note below

partial pressure of hydrogen sulfide does not exceed 0.3 psia and its hardness is not greater than 36HRC.

4.13.1.6   Non-ferrous Metals   Most of the common non-ferrous metals and alloys are acceptable provided they meet the hardness criteria as mentioned in Table 4.13.

Specific guidelines exist for titanium alloys. Hydrogen embrittlement of titanium alloys may occur if galvanically coupled to certain active metals (i.e., carbon steels). Some titanium alloys may be susceptible to crevice corrosion and chloride SCC. Hardness has not been shown to correlate with susceptibility to SSC in titanium alloys.

4.13.1.7   Other Aspects   In addition to Materials, the Standard gives guidelines on other aspects related to materials. These aspects are listed below:

- Fabrication: Overlays, Welding, Identification Stamping, Threading and Cold-Deformation Processes
- Bolting: Both Exposed and Non-exposed
- Coating and Plating
- Special Components: Bearings, Springs, Instrumentation Devices, Seal Rings
- Valves and Chokes
- Wells, Flow Lines, and other Production Facilities.

For specific requirements the Standard must be looked into in a very detailed manner.

REFERENCES

Bates J. F., (1968), Sulphide Stress Corrosion Cracking of High Strength Seels in Sour Crude Oils. Proceedings of 24th Conference on National Association of Corrosion Engineers, March 18–22; Clevaland, OH, p. 370.

Bobby Kannan, M., Raja, V. S., Raman, R., and Mukhopadhyay, A .K., (2003), Influence of Multi-Step Aging on the Stress Corrosion Cracking Behavior of 7010 Al Alloy. Corrosion. 59, pp. 881–889.

Bobby Kannan, M., Raja, V. S., Raman, R., Mukhopadhyay, A. K. and Schmuki, P., (2005), Environmental Assisted Cracking Behavior of Peak Aged 7010 Aluminum Alloy Containing Scandium. Metall. Mater, Trans., 36A, pp. 3257–3262.

Chawala, S. L., and Gupta, R. K., (1993), *Materials Selection for Corrosion Control*, ASM International, Ohio, p 305.

Copson, H. R. (1959), Physical Metallurgy of Stress Corrosion Fracture, Interscience, New York, p 247.

Craig, B. D., (1989), (Ed) *Handbook of Corrosion Data*, ASM International, Ohio.

Davis, J. R., (1993), (Ed) *Aluminum and Aluminum Alloys*, ASM International, Ohio.

Davis, J. R., (1999), (Ed) *Corrosion of Aluminum and Aluminum Alloys*, ASM International, Ohio.

De Renzo, D. J., (1985), *Corrosion Resistant Materials Handbook*, 4th Edition, Noyes Data Corp, New Jersey.

Garner, A., (1983), *Pitting Corrosion of High Alloy Stainless Steel Weldment in Oxidizing Environment* Weld. J., 62 (1), p 29.

Graville, B. A., (1976), Cold Cracking in Welds in HSLA Steels, Welding of HSLA (microalloyed) Structural Steels. Proceedings of International Conference on American Society for Metals, November.

Kreysa, G., and Schutze, M., (2005), (Eds) *Corrosion Handbook: Corrosive Agents and Their Interaction with Materials* (1–16) 2nd Edition, Wiley-VCH, Frankfurt.

McCauley, R.A., (1994), *Corrosion of Ceramic and Composite Materials*, 2nd Edition, Marcel Dekker Inc, NY.

McEvily, A. J., (1990), *Atlas of Stress-Corrosion and Corrosion Fatigue Curves*, ASM International, Ohio.

Munger, C., (1984), *Corrosion Prevention by Protective Coatings*, NACE international Publication, Houston.

Richlen, S. L., and Parks, W. P., (1991), Heat Exchangers, Engineered Materials Handbook, Vol 4, *Ceramics and Glasses*, ASM International, Ohio.

Rotheiser, J. I., (2000), Design of plastic products, *Modern Plastics Handbook/Modern Plastics*, C. A. Harper (ed). McGraw-Hill, New York, p 8.1.

Schweitzer, P. A., (1995), *Corrosion Resistance Tables: Metals, Nonmetals, Coatings, Mortars, Plastics, Elastomers and Linings and Fabrics*, 4th Edition, Marcel Dekker, New York.

Sreedhar, G., and Raja, V. S., (2010), Hot Corrosion Study of the YSZ/Al_2O_3 Dispersed NiCrAlY Plasma Sprayed Coatings in Na_2SO_4 – 10 wt.% NaCl Melt, Corros. Sci., 52 pp. 2592–2602.

Speidel M. O (1981) Stress Corrosion Cracking of Stainless Steels in NaCI Solutions, Metallurgical Transactions A, Vol. 12a, p. 779.

Wernick, S., Pinner, R., Sheasby, P. G., (1987), *Surface Treatment and Finishing of Aluminum, and Its Alloys*, 5th edition, ASM, Metals Park.

Wilks, E. S., (2001) (Ed) *Industrial Polymers Handbook: Products, Processes, Applications*, Wiley-VCH, Weinheim.

SOME IMPORTANT NACE STANDARDS

NACE Standard RP0403-2003, Avoiding Caustic Stress Corrosion Cracking of Carbon Steel Refinery Equipment and Piping.

NACE Standard RP0403-2003, Avoiding Caustic Stress Corrosion Cracking of Carbon Steel Refinery Equipment and Piping.

NACE Standard TM0284-2003, Evaluation of Pipeline and Pressure Vessel Steels for Resistance to Hydrogen-Induced Cracking.

NACE Standard TM0103-2003, Laboratory Test Procedures for Evaluation of SOHIC Resistance of Plate Steels Used in Wet H_2S Service.

ANSI/NACE Standard TM0177-96, Laboratory Testing of Metals for Resistance to Sulfide Stress Cracking and Stress Corrosion Cracking in H2S Environments.

NACE Standard RP0472-2000, Methods and Controls to Prevent In-Service Environmental Cracking of Carbon Steel Weldments in Corrosive Petroleum Refining Environments.

NACE Standard RP0296-2004, Guidelines for Detection, Repair, and Mitigation of Cracking of Existing Petroleum Refinery Pressure Vessels in Wet H_2S Environments.

NACE Standard RP0170-2004, Protection of Austenitic Stainless Steels and Other Austenitic Alloys from Polythionic Acid Stress Corrosion Cracking During Shutdown of Refinery Equipment

NACE MR0175/ISO 15156-1:2001(E) Petroleum and Natural Gas Industries— Materials for use in H2S-Containing Environments in Oil and Gas Production.

NACE Standard MR0103-2003: Materials Resistant to Sulfide Stress Cracking in Corrosive Petroleum Refining Environments.

5

FAILURE ANALYSIS PROCEDURE WITH REFERENCE TO CORROSION FAILURES

5.1 INTRODUCTION

As mentioned in the introductory chapter of this book, corrosion is one of the deteriorating modes of metallic equipments (among overload fracture, fatigue, creep, wear, etc.) leading to either "slow deterioration" in functioning (thickness reduction, insulating scale build-up, choking, etc.) or unexpected "sudden" failures like stress corrosion cracking, hydrogen cracking, etc. Hence, all failures are not due to corrosion and all corrosion phenomena are not failures. Unless one conducts a systematic exercise that probes into all aspects of the experienced failure, one cannot, and should not, conclude about the mode of the failure. This analysis is called *Failure Analysis* and is very relevant in a book dealing with *Corrosion Failures*. Carrying out detailed analysis of every unexpected failure is a must to ensure the basic cause of the failure and to take preventive actions so that similar failures do not occur in the future. Just replacement alone would not guarantee against the future failure. On the other hand, immediate replacement is a must and is an economical necessity.

The procedure employed in general analysis of failures of equipments/components in chemical process industries occurred during service would be briefly covered in this chapter along with precautions to be taken during the analysis, with frequent and specific reference to corrosion failures.

Corrosion Failures: Theory, Case Studies, and Solutions, First Edition. K. Elayaperumal and V. S. Raja.
© 2015 John Wiley & Sons, Inc. Published 2015 by John Wiley & Sons, Inc.

The purpose of failure analysis, the various stages involved along with the different techniques employed, analysis of the observations and test results, diagnosis of the failure, making recommendations for remedial measures, and finally preparation of the failure analysis report would be briefly covered in this chapter. Following these procedures will enable not only in identifying one or more of the corrosion mechanisms discussed in Chapter 2, but also the causating factor for such corrosion. For detailed procedures, the readers can refer the ASM Handbook on "Failure Analysis and Prevention" (2002).

5.2   PURPOSE OF FAILURE ANALYSIS INVESTIGATIONS

A failure investigation is aimed at determining the cause of the unexpected failure which occurred in service. Failure of equipment is broadly defined as unexpected losing of its functionality during service. The primary cause may be related to design deficiency, material defect, mechanical forces or environmental interactions, and inadequate control of process parameters and maintenance. Corrosion is one of the prominent most common environmental interactions, particularly in chemical process equipments.

The purpose of the investigation is to arrive at the specific cause which has led to the failure and to make recommendations for remedial actions. As far as corrosion failures are concerned, the investigations should include techniques to arrive at one of the different Forms of Corrosion, which were discussed earlier in Chapter 3 of this book.

5.3   FAILURE ANALYSIS STEPS

The detailed failure analysis involves roughly the following steps.

5.3.1   Site Visit

Site visit forms the first major step in the failure analysis exercise, after getting briefed over correspondence about the broad generic nature of the failure. It is needless to emphasize the importance of the site visit. The main purpose of the site visit is to get first hand information on the design, material of construction (MOC), specified and operating parameters, the failure incident, and operating history, to physically see the failed component/equipment/location, etc. and to formulate and get approved the detailed failure analysis procedure before starting the latter.

Going by the experience of the authors, any short cut to avoid the site visit, due to time, cost, or any other reason, is likely to result in wrong interpretation, wrong analysis, and wrong remedial measures which would lead to major multiple failures. There was a case of bad experience of receiving a sample for failure analysis over a courier service to avoid a "time consuming expensive site visit." It was realized that the sample was cut from a high pressure pipeline without paying careful attention to the instructions given by the analyst regarding (i) details of orientation of the needed sample, (ii) cutting procedure, (iii) location of the sample, etc. The whole courier exercise resulted not only in destroying the critical finger print of the failure, but also further time lost and replacement of the pipeline without arriving at the real cause of failure.

The task to be carried out at site

- Obtain background data regarding design, function, MOC, service conditions, past inspection records, the subject failure incident, and past history of similar or related incidents. If the equipment was exposed to corrosive environment in service, as in chemical process industries, the service conditions should include detailed chemical composition of the fluids involved. If the equipment concerned is a heat exchanger, shell, and tube or plate type, both side (cooling side and heating side) fluid compositions and the respective service conditions should be obtained. Variations beyond design conditions, start-up, up-sets, and shut-down conditions over a meaningful period of time prior to the failure should also be noted.
- Inspect the failed equipment and also the nearby upstream and downstream equipments to the extent accessible to get related comparable information.
- Closely examine the failed area, failed spot on the equipment, and record the features photographically and graphically. Macroscopically examine and analyze fracture surfaces, secondary cracks, and other external surface features. One should also look for characteristic sites of initiation of cracks, features of the initiation sites, and whether cracks have single origin site or multiple origin sites. One should also look for the presence of fatigue striations which may be present on samples subjected to cyclic stresses.
- Plan the investigative procedure based on the above information at the site itself before proceeding further. The plan would decide the location, orientation, and size of the representative samples to be cut.
- Obtain representative cut samples containing the failed spots and the failure features and also those from typical unfailed areas if sample cutting becomes essential. Cutting of samples may not be possible and/or may not be necessary in many cases. Detailed record of the appearance of the failure has to be relied upon in such cases. At times such record is

itself sufficient. If necessary, non-destructive-tests such as radiography, ultrasonic testing, dye-penetrant testing, in-situ metallography, etc. need to be carried out on the equipment in position at the failed locations.

- Obtain representative samples of scales, deposits, corrosion products, etc. in loose/adherent contact with the inside surface of the equipment on the process side and also those on the outside surface if external corrosion is suspected.
- Ensure that the samples are well-preserved. They should be properly identified, photographed, degreased, and cleaned with detergents (not acidic ones) dried and preserved in dry packets and stored under dry conditions before taking for detailed tests.

5.3.2   Tests on the Samples

The purpose of the tests is

- to check the nature and purity of the metal and deposit samples,
- to detect if any unusual impurity has been present in the medium,
- to check conformity of the equipment and the material of construction with the stated specification under which it was designed, fabricated, and put to use,
- to trace out the progress of the failure mode and finally,
- to diagnose the observed unexpected problem (purpose of this exercise).

Importance/Purpose of Sample Testing:

It is not always necessary to carry out sample testing. At times, it is possible to diagnose the failure by close visual examination alone with or without visual aids like magnifying lens, dye-penetrant tests, boroscope, etc. and also using nondestructive testing (NDT) techniques (Raj et al. 1999) like ultrasonic technique and radiography technique. Detailed destructive sample testing becomes essential when such a diagnosis based on visual/NDT examination alone is not possible. For example, a visual crack may be pure mechanical crack, a fatigue crack, a stress corrosion crack, etc. Unless detailed microscopic examination on a properly cut sample is carried out, one cannot short list the above. Similarly an unusually high corrosion rate could be due to abnormalities in chemical composition of the MOC. Such a possibility can be confirmed only after a detailed chemical analysis of a properly cut sample. In view of these necessities sample testing becomes a must.

The tests have to be formulated keeping in mind that the purpose is to quickly arrive at the diagnosis of the specific problem faced and not to publish an extended research paper. The latter, if scope shows itself as the result of the diagnosis, could be taken up as a long-term research program if necessary and feasible.

Some of the tests are listed below. All of these need not be carried out. Depending upon the merit of each case the tests have to be chosen:

- Nondestructive tests before cutting/drilling the sample for further tests: Radiography and/or ultrasonic crack detection to detect any subsurface cracks induced by phenomena such as hydrogen cracking and blistering. Thickness measurement to assess any loss of metal by general uniform corrosion. Dye-penetrant tests to detect very fine surface cracks not visible to the naked eye during visual examination. Hardness tests on welds, heat-affected zones, parent metal portions, and also on portions near cracked areas.

- Fractographic examination with recording of the fracture surface before cutting the sample for destructive examination. This would be useful to classify the fracture as ductile or brittle, intergranular or transgranular, stress corrosion cracking, or corrosion fatigue, etc. Preliminary Fractographic examination can be done with a simple Stereomicroscope to obtain relatively macro features. However, for a detailed analysis of fractured surface, scanning electron microscopy (SEM) needs to be used. For those specimens covered with corrosion products, appropriate cleaning measures (ultrasonic, chemical cleaning) need to be taken.

- Chemical analysis of the bulk sample to check conformity with specifications. Any major deviation from the specification limits should be rechecked. Chemical analysis should also be carried out on surface corrosion products, on deposits, and on coatings. Service conditions and fluid composition would indicate the elements to look for in corrosion products: Fe, Cr, Ni, Mo, S, Si, Mn, etc. Techniques such as Atomic emission spectroscopy, Atomic absorption spectroscopy, and X-ray fluorescence spectroscopy are the commonly used techniques. It should be emphasized that appropriate test standards are required to obtain accurate data.

- X-ray diffraction and electron probe micro analysis (EPMA) are carried out on different areas of the metal surface and on corrosion products, scales, and deposits for detecting inorganic products such as oxides, chlorides, sulfides, etc. While X-ray diffraction provides information on phases present, EPMA is used to obtain chemical composition at the localized regions of the samples. If one desires to know the oxidation states of the corrosion products, X-ray photo electron spectroscopy can be used.

- Mechanical tests.

These include tensile tests, hardness tests, and impact tests. These tests are necessary to assess cases involving premature cracks like stress corrosion cracking, corrosion fatigue cracks, and hydrogen cracks. Conformity of the original material used for the equipment with respect to mechanical property requirements is checked by these tests. The ductility and hardness values would be also useful to detect any environment-induced embrittlement during service.

- Microscopic examination: In addition to the scanning electron microscopic examination carried out on the bare fracture surface, optical microscopy should be carried out on the cross sections of properly cut and metallographically polished and etched samples. This test would provide a whole variety of information about all localized corrosion phenomena discussed in Chapter 3. Metallurgical surface phenomena such as oxidation, sulfidation, carburization, nitriding, decarburization, etc. and their extent across the thickness, pitting, dealloying (dezincification), erosion indications, occurrence of sensitization in stainless steels leading to intergranular corrosion, crack propagation mode in stress corrosion cracking whether transgranular or intergranular or mixed, and presence of corrosion fatigue cracks are some of the features detectable through optical metallography. On heat exchanger tubes, leakage through fine pin-hole pits is due to localized corrosion starting either on the inside (tube side) or outside (shell side) surface. A cross sectional optical microscopic examination would throw light on this. Accordingly further investigations can be concentrated either on the tube side or on the shell side.

It is necessary to emphasize the fact that the above tests are to be done with utmost care and the user should also realize the limitations of each of the equipment used for analysis. For a more detailed study on materials characterization techniques, readers can go through the book Materials Characterization (1992).

5.3.3 Analysis, Interpretation, and Diagnosis of the Failure

The site observations and sample test results should be analyzed together as a whole package with the sole purpose of arriving at the diagnosis of the failure. If necessary, support from published literature should be obtained. All these should be viewed together with the aim of arriving at the right diagnosis, the root cause of the failure.

Possibly different hypotheses for the diagnosis of the failure have to be considered. Based on the site observations and test results, the weak hypotheses have to be eliminated one-by-one. For example, a sudden failure by fracture could be a brittle fracture due to purely mechanical overload, stress corrosion cracking, to corrosion fatigue or due to stress concentration at the bottom of a sharp corrosion pit, or a ductile fracture as a result of wall thinning due to general corrosion or intergranular corrosion. SEM and/or optical microscopic results would help to pinpoint the right one among the possibilities. Further, a highly localized corrosion could be a straight chloride pitting corrosion on a smooth surface or a crevice corrosion in a geometrical discontinuity or under a deposit. Site discussion with the concerned personnel on design, on chloride concentration and temperature, and on the possibility of porous deposit formation would help to choose the right one among the above.

A quick literature search for published reports/papers on similar incidences in the past and for theoretical support for the arrived diagnosis would be very helpful towards the confirmation of the diagnosis. For literature search, it should be kept in mind that the search is highly focused, very specific, practical-oriented, and to be completed within a very short time. Such literature supports would be very welcome among the practicing plant personnel.

5.4   FAILURE ANALYSIS REPORT: CONTENTS AND PREPARATION

The failure analysis report should be written in a simple language, clearly, consciously, and logically. It should be easily understandable by a practicing nonspecialist, neither a corrosion chemist, engineer, nor a metallurgist, for corrosion failures. Too much of scientific supports are not necessary, they would mask the specific diagnosis in a sea of information.

Specifically, the report should contain the following.

- Necessity of carrying out the said failure analysis, justifying that the failure is premature.
- Brief description of the chemical process involved.
- Description of the failed component and the failure incident giving the prior chronological happenings, prior service history.
- Actual service condition at the time of failure, highlighting any deviations from designed parameters.
- The used material of construction of the equipment/component, its specification, its manufacturing sequence, and metallurgical treatment given to it during manufacture.
- Factual summary of the site observations and findings obtained during discussion with the plant personnel.

- The actual results of all the tests conducted on the cut and obtained samples, both the failed one and the non-failed baseline sample, if tests were conducted on the latter.
- Specific mention about the metallurgical quality of the component as seen through the test results.
- Interpretation, discussion, and analysis of all the input information.
- Specific diagnosis of the failure, if necessary stage by stage mechanisms/ phenomena that caused the ultimate failure.
- Explanation of all the observed symptoms using the stated diagnosis.
- Clear conclusions on the causes of failure.
- Easily implementable practical recommendations towards avoidance of repetition of similar failures in the particular site. The recommendations should be very concise and specific and should be implementable within a relatively short time.
- Towards avoiding corrosion phenomena, if any superior material of construction is specified (one of the corrosion resistant alloys) and if quick coupon tests in the running plant are desired, the recommendations should include the procedural details of such coupon tests.
- Finally, the report must contain an *Executive Summary*, pasted in the beginning, highlighting the *Diagnosis and Recommendations*.

REFERENCES

ASM Handbook, (1992), *Materials Characterization*, Vol. 10, 9th Edition, ASM Handbook, OH, USA.

ASM Handbook, (2002), *Failure Analysis and Prevention*, Vol. 11, 10th Edition, ASM International, Materials Park, OH.

Raj, B., Jayakumar, T., Thavasimuthu, M., (1999), *Practical Nondestructive Testing*, Narosa Publications, New Delhi.

6

CASE STUDIES

6.1 PREAMBLE

The book so far has dealt with the Basics of Corrosion, Cost of Corrosion, Forms of Corrosion, Materials of Construction, and Procedures of Failure Analysis of Corrosion Failures. The treatments in each chapter were not scientifically elaborate, but were aimed to present the subjects in a brief practical manner so that the readers are somewhat prepared in going through specific case studies with ease. It is the desire of the authors that the readers, particularly the plant engineers who are getting initiated into corrosion solving in the plants, read these chapters before looking into the specific case studies of their interest. What follows is the presentation of the Case Studies themselves on case to case basis without linking the cases among themselves.

The case studies arose from authors' own investigations/studies of unexpected corrosion failures of process equipments. These problems occurred in a variety of chemical process industries. These include Oil and Gas Industries, Oil Refineries, Captive Power Stations, Fertilizers, Petrochemicals, Organic Chemicals, Inorganic Chemicals, Textile Dyes, etc., including Tube Manufacturing Industries. The readers are not expected to read through all of them at one stretch. In order to help them to choose particular case study of individual's interest, the case studies are classified with respect to Industry-wise,

Corrosion Failures: Theory, Case Studies, and Solutions, First Edition. K. Elayaperumal and V. S. Raja.
© 2015 John Wiley & Sons, Inc. Published 2015 by John Wiley & Sons, Inc.

Corrosion Form-wise, Material of Construction-wise, Diagnosis-wise, and Remedial measure-wise, numbered and indexed. This would facilitate the reader to choose the particular case study of interest.

Attempt has been made to include brief descriptions of all the forms of corrosion mentioned under the heading of "Diagnosis" in the earlier Chapter on Forms of Corrosion. The readers are advised to refer to this chapter whenever some exposure and/or clarifications are needed while going through the specific case studies.

CLASSIFICATION OF CASE STUDIES

Industry-wise Classification of Case Studies of Corrosion Failures

Sr. No	Industry Type	Case Study Numbers
1	Inorganic chemicals	1, 5, 7, 22, 32, 33, 47, 50, 51, 56, 67, 80
2	Organic chemicals	1, 8, 10, 11, 12, 14, 18, 19, 21, 35, 40, 49, 58, 65, 68, 69, 74, 75, 76,
3	Oil refinery	3, 61, 63, 64, 78, 79
4	Captive power plants, package boilers,	4, 17, 38, 54, 57, 66, 72
5	Oil and gas separation plants	6, 16
6	Pulp and paper plants	9, 70
7	Petrochemical plants	13, 20, 24, 28, 30, 31, 37, 44, 46, 54, 55, 73
8	Fertilizer plants	15, 25, 26, 29, 36, 42, 45, 59, 62, 69, 71, 77
9	Air conditioning/refrigeration units	23, 34, 53
10	Textile dyeing units	27, 48, 52, 60
11	Storage water heater	39
12	Stainless steel tube making plant	41
13	Hospital waste treatment unit	43

Material-wise Classification of Case Studies of Corrosion Failures

Sr. No	Material Type	Case Study Numbers
1	Carbon steels	2, 4, 5, 12, 14, 17, 21, 26, 30, 35, 38, 43, 44, 46, 49, 50, 56, 57, 59, 63, 64, 66, 71, 74, 75, 77, 78
2	Stainless steels	5, 6, 8, 9, 10, 11, 13, 16, 18, 19, 20, 22, 23, 24, 25, 27, 28, 29, 31, 32, 33, 39, 40, 41, 44, 45, 48, 50, 51, 52, 60, 61, 62, 65, 67, 68, 69, 70, 73, 76, 80
3	Alloy steels	36, 42, 44, 69, 79
4	70/30 nickel copper alloy	2, 47, 58
5	90/10 copper nickel alloy	3, 34
6	Alloy 800H(T)	15,37
7	Nickel-molybdenum alloy C-276	7
8	Copper	53
9	Admiralty brass	54
10	Cast iron	72
11	Titanium	55

Diagnosis-wise Classification of Case Studies of Corrosion Failures

Sr. No	Diagnosis	Case Study Numbers
1	General uniform corrosion	1, 2, 3, 23, 31, 32, 37, 39, 41, 43, 44, 48, 49, 50, 53, 58, 59, 67, 68, 77, 78, 79
2	Pitting corrosion	14, 24, 26, 30, 38, 48, 52, 59, 66, 71, 73, 74
3	Stress corrosion cracking by chloride	11, 13,19, 20,31, 33, 48, 50, 58, 60, 61, 62
4	Erosion corrosion	5, 34, 35, 40, 72, 75, 76
5	Galvanic corrosion	16, 19,34, 44, 47

(continued)

Sr. No	Diagnosis	Case Study Numbers
6	Intergranular corrosion	8, 17, 25, 28, 45, 70
7	Stress corrosion cracking by caustic	9, 18, 63,80
8	Crevice corrosion	6, 27, 29, 55, 65
9	Caustic gouging	21, 42,64
10	Micro-biological corrosion	7, 46, 53
11	Carburization	15, 69
12	Hot wall effect corrosion	3
13	Hot ash corrosion	4
14	Stress corrosion cracking by polythionic acid	10
15	Stray current corrosion	12
16	Corrosion fatigue	18
17	Sulfidation	36
18	Hydrogen attack	51
19	Dezincification	54
20	Dew point corrosion	57
21	High temperature attack	56

Remedy-wise Classification of Case Studies of Corrosion Failures

Sr. No	Remedy Type	Case Study Numbers
1	Design	3, 6, 12, 17, 18, 20, 22, 35, 42, 45, 56, 57, 61, 65, 67, 73, 75, 76
2	Operation	4, 10, 11, 13, 14, 17, 20, 21, 22, 23, 26, 27, 30, 31, 36, 37, 38, 41, 43, 44, 46, 48, 51, 52, 55, 57, 58, 60, 62, 63, 66, 68, 77, 78, 79

Sr. No	Remedy Type	Case Study Numbers
3	Selection of material of construction	1, 2, 9, 11, 13, 18, 19, 24, 33, 36, 40, 45, 49, 50, 51, 54, 56, 58, 59, 60, 62, 69, 70, 75, 77, 78, 80
4	Fabrication	8, 11, 19, 29, 32, 33, 39, 63, 80
5	Quality control	5, 15, 16, 22, 25, 28, 34
6	Water treatment	7, 53, 59, 64, 74
7	Cathodic protection/coating	71, 72

GENERAL

Loss due to Corrosion in Chemical Process Industries

SERVICE	Many services involving equipments in chemical processes with varying media like acids, alkalis, elevated temperatures, etc.
PROBLEM	Unexpected thinning, pin holes, pitting, cracking, etc. leading to leakage losses, equipment losses, environmental contamination, and even human losses
MATERIAL	Carbon steels, stainless steels, copper base alloys, aluminium alloys, nickel base alloys, and titanium
OBSERVATIONS	Both general uniform corrosion and all forms of localized corrosion like pitting, galvanic corrosion, crevice corrosion, intergranular corrosion, stress corrosion cracking, erosion corrosion, etc.
DIAGNOSIS	1. Operation beyond design parameters 2. Non-conformance to quality of equipments 3. Un-traced impurities in the process media 4. Wrong materials-of-construction 5. Wrong design 6. Non-monitoring of corrosion processes
REMEDY	Creation of CORROSION AWARENESS and follow-up action

CASE STUDY 1   Bromine Preheater in a Pharmaceutical Fine Chemical Plant

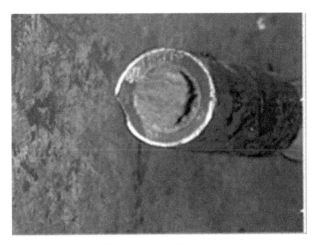

Case 1   Heavy corrosion and scaling in Monel-400 bromine preheater.
(Courtesy: Elayaperumal, in "Corrosion Failures in Process Industries in India: Statistical Analysis and Case Studies" in Special Supplement 2009, Corrosion Reviews Special Issue: India, England: Freund Publications, 2009, p 51, case study 1.)

SERVICE	Horizontal preheater coil, for electrical heating bromine vapor from 80 to 360°C prior to an organic reaction
PROBLEM	Leakage within 9 months of operation, first near the high temperature end subsequently in other places
MATERIAL	70/30 NICKEL COPPER ALLOY (MONEL 400) pipe of size 40 mm OD × 3.8 WT
OBSERVATIONS	1. Heavy corrosion/scaling on the inside surfaces, resulting in thickness reduction 2. Preferential corrosion at the 6 o'clock position 3. Leaks at the bottom half portions (between 4 and 8 o'clock positions)
DIAGNOSIS	HIGH UNIFORM CORROSION (at a rate of about 5.1 mm/year) by moist bromine liquid saturated with oxygen. Monel is resistant only to dry bromine free of moisture and dissolved oxygen
REMEDY	Use Nickel-200 in the place of Monel-400. Ensure bromine is fully vaporized before it reaches the preheater. Avoid moisture and air contamination

CASE STUDY 2 Structurals in a White Clay Manufacturing Plant	
IRON CONTAMINATION OF WHITE CLAY POWDER, MAKING THE LATTER UNUSABLE	
SERVICE	White clay manufacturing involves handling of clay slurry through bleaching, filtering, and conveying the filter cake. INORGANIC CHEMICAL PLANT
PROBLEM	Iron oxide (rust) flakes liberated due to corrosion of structures associated with filter presses and conveyors, fall into the process media, wet cake, and contaminate the product
MATERIAL	CARBON STEEL plates
OBSERVATIONS	Peeling-off of the paint coating followed by corrosion of the underlying steel resulting in brownish rust flakes. The latter could be scrapped-off easily from the substrate steel surface
DIAGNOSIS	GENERAL UNIFORM CORROSION by the process aqueous solution of pH 4.5/5.0, at a very slow rate of about 0.025 mm/year. This rate is quite acceptable from structural integrity point, but not from product purity point for the specific application
REMEDY	Replace the structures with solid FRP or with galvanized steel

(Courtesy: Elayaperumal, in "Corrosion Failures in Process Industries in India: Statistical Analysis and Case Studies" in Special Supplement 2009, Corrosion Reviews Special Issue: India, England: Freund Publications, 2009, p 50, case study 2.)

CASE STUDY 3   Sea Water Cooler Tubes in an Oil Refinery

Case 3   Hot-wall effect corrosion by sea water on cupro-nickel cooler tube in an oil refinery. (Courtesy: Elayaperumal, in "Corrosion Failures in Process Industries in India: Statistical Analysis and Case Studies" in Special Supplement 2009, Corrosion Reviews Special Issue: India, England: Freund Publications, 2009, p 52, case study 3.)

SERVICE	Cooler tubes of a horizontal shell-and-tube cooler in an oil refinery. Shell side: hot hydro carbon gas from catalytic reforming unit. Tube side: sea water for cooling
PROBLEM	Leaks within a few months in top row of tubes on the final pass of the sea water very close to the shell side inlet nozzle. Hottest portion on both shell side and tube side of the tubes
MATERIAL	90/10 CUPRO-NICKEL
OBSERVATIONS	1. General corrosion by sea water on the inside surface only on the portion corresponding to the hot gas entry on the shell side 2. Very clear demarcation line between the affected and unaffected portions 3. Only the top rows are affected
DIAGNOSIS	HOT-WALL EFFECT CORROSION by sea water. Happens only where a localized heat input is there on the shell side
REMEDY	Use a perforated impingement plate below the shell side inlet nozzle so that the heat gets distributed

CASE STUDY 4   Package Boiler Tube in an Organic Chemical Plant

Case 4   Corrosion by fuel-ash deposit in a package boiler tube.
(Courtesy: Elayaperumal, in "Corrosion Failures in Process Industries in India: Statistical Analysis and Case Studies" in Special Supplement 2009, Corrosion Reviews Special Issue: India, England: Freund Publications, 2009, p 53, case study 4.)

SERVICE	Package boiler, water wall tube, coal/oil fired on the shell side
PROBLEM	Leak after 22 years of service in 25 out of 900 tubes. Interior not easily accessible tubes only affected
MATERIAL	BOILER QUALITY CARBON STEEL
OBSERVATIONS	Corrosive attack at localized places initiated on the shell side surface under fuel-ash deposit. Insufficient soot blowing over the years
DIAGNOSIS	CORROSION BY LOW MELTING EUTECTICS of vanadium, sodium, and sulfur compounds originating from the fuel
REMEDY	1. Periodic cleaning of ash deposits by high pressure jet of air, steam, or water 2. Maintenance and operation of soot blowers in working condition 3. Dose conventional fuel additive chemicals into the fuel, particularly the oil fuel

CASE STUDY 5 Shell of a Packed Column for Ammonia and Water Contact in an Ammonia Processing Plant	
HEAVY WALL THICKNESS REDUCTION ON LOCALIZED AREA IN THE SHELL	
SERVICE	Contact between upward ammonia gas with downward flow of water, through packing of corrugated metallic bricks
PROBLEM	Heavy wall thickness reduction of the shell on certain vertical portions within 16 months of service
MATERIAL	Shell: CARBON STEEL; Packing: TYPE 316 STAINLESS STEEL
OBSERVATIONS	1. Vertical groove type attack on the inside surface of the shell where thickness was very low 2. One of the stainless steel packing bricks had completely thinned down and disintegrated into sharp pieces 3. Chemical analysis of the disintegrated brick did not correspond to 316 SS, not even to 304 SS, but to lower quality non-standard composition
DIAGNOSIS	SOLID PARTICLE SLURRY EROSION of the shell by the disintegrated sharp pieces of the specific packing which was poor in quality
REMEDY	Ensure quality of the packing to conform to design specifications

(Courtesy: Elayaperumal, in "Corrosion Failures in Process Industries in India: Statistical Analysis and Case Studies" in Special Supplement 2009, Corrosion Reviews Special Issue: India, England: Freund Publications, 2009, p 54, case study 5.)

CASE STUDY 6   Instrumentation Tube in an Offshore Platform of an Oil and Gas Plant

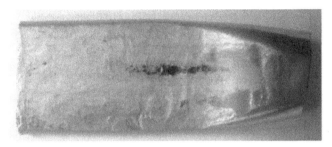

Case 6   Crevice corrosion in an off-shore platform stainless steel instrumentation tube. (Courtesy: Elayaperumal, in "Corrosion Failures in Process Industries in India: Statistical Analysis and Case Studies" in Special Supplement 2009, Corrosion Reviews Special Issue: India, England: Freund Publications, 2009, p 55, case study 6.)

SERVICE	Instrumentation tube in an off-shore platform producing oil and gas. Contains produced gas under pressure at ambient temperature
PROBLEM	Minor leak after 5 years of operation under one of the support clamps
MATERIAL	TYPE 316 L STAINLESS STEEL 3/8″ OD × 0.065″ WT seamless as per A268
OBSERVATIONS	Highly localized corrosion on the outside surface of the tube below the tight fit plastic support clamp. No general corrosion elsewhere due to the overall marine atmosphere. No corrosion on the inside surface
DIAGNOSIS	CREVICE CORROSION under the tight fit plastic support clamp
REMEDY	Keep the plastic support clamps under loose fit so that free air and moisture flow during winds are ensured

CASE STUDY 7   Plate Type Heat Exchanger/Cooler in a Sulfuric Acid Plant

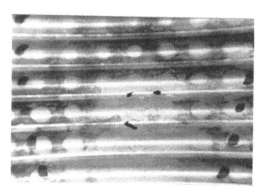

Case 7   Microbial corrosion on Hastelloy C-276 plate- type heat exchanger/cooler (PHE) element in a sulfuric acid plant.
(Courtesy: Elayaperumal, in "Corrosion Failures in Process Industries in India: Statistical Analysis and Case Studies" in Special Supplement 2009, Corrosion Reviews Special Issue: India, England: Freund Publications, 2009, p 57, case study 7.)

COOLING WATER SIDE OF THE AFFECTED PLATE ELEMENT

SERVICE	In Sulfuric acid plants, the final product is cooled through a PHE (plate type heat exchanger). On one side of the plate elements acid enters at 90°C and cools to 70°C. On the other side of the element cooling water flows
PROBLEM	Development of pin holes in the elements within a year of operation resulting in leakage, allowing water to enter the acid
MATERIAL	HASTELLOY C-276 (UNS N10276)
OBSERVATIONS	Sharp holes on the acid side of the elements with corrosive attack (etched surface) downstream of the holes. Shiny central spots could be seen on the water side corresponding to these holes
DIAGNOSIS	MICRO-BIOLOGICAL CORROSION by living organisms, bacteria present in the water. Bacteria strongly cling to the metal surface and give rise to pinhole attack under tiny colonies
REMEDY	Kill the microbial organisms by adding biocides, both gaseous chlorination and addition of proprietary chemicals to the cooling water

CASE STUDY 8   Dissimilar Stainless Steel Weld in an Organic Chemical Plant

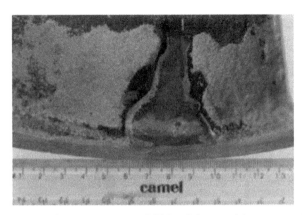

Case 8   Galvanic corrosion on parent metal 304 stainless steel in contact with weld metal 316 stainless steel in an organic chemical plant.
(Courtesy: Elayaperumal, in "Corrosion Failures in Process Industries in India: Statistical Analysis and Case Studies" in Special Supplement 2009, Corrosion Reviews Special Issue: India, England: Freund Publications, 2009, p 57, case study 8.)

SERVICE	A pipe-flange assembly both having longitudinal welds processing an organic amine chloride solution
PROBLEM	Preferential corrosion by the process solution on the weld and surrounding areas resulting leakage
MATERIAL	Pipe and flange: TYPE 304 STAINLESS STEEL Weld metal: TYPE 316 STAINLESS STEEL
OBSERVATIONS	1. Inner surface showed heavy corrosion on the weld parent metal boundary area and also on nearby areas on the parent metal and the weld metal on either side of the weld 2. Galvanic corrosion of parent metal. Metallography indicated inter-granular corrosion (IGC)
DIAGNOSIS	INTERGRANULAR CORROSION in the heat-affected zone (HAZ) of the parent metal on either side of the weld
REMEDY	For welding 304 SS pipes use 304L SS consumables and weld with lowest possible heat input to minimize HAZ area

CASE STUDY 9   Digestor Preheater in a Pulp and Paper Plant

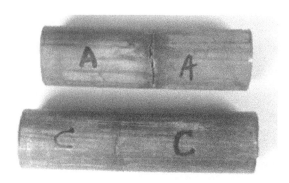

Case 9   Caustic stress corrosion cracking in stainless steel digestor pre-heater in a pulp and paper plant.
(Courtesy: Elayaperumal, in "Corrosion Failures in Process Industries in India: Statistical Analysis and Case Studies" in Special Supplement 2009, Corrosion Reviews Special Issue: India, England: Freund Publications, 2009, p 58, case study 9.)

SERVICE	Vertical shell-and-tube heat exchanger. Shell side: steam. Tube side: digestor caustic liquor heated to 170°C and fed to the digestor
PROBLEM	Leakage within 7 months of operation. Four identical ones behaved in the same way
MATERIAL	TYPE 316 L STAINLESS STEEL seamless tubes of size 32 mm OD × 1.50 mm WT, 5.0 m long
OBSERVATIONS	1. Circumferential cracks somewhere in the middle portion between two baffle positions 2. Some tubes had bends in the middle 3. Multiple cracks initiated on the ID side, digestor liquor side, trans-granular and branched during propagation
DIAGNOSIS	TRANS-GRANULAR STRESS CORROSION CRACKING BY CAUSTIC on the stress areas between the baffles
REMEDY	Replace 316L SS with 316 SS with highest possible yield strength and wall thickness. Redesign the baffle position and numbers so that stress on the tubes is minimized

CASE STUDY 10   Esterification Column in an Organic Chemical Plant

Case 10   Transgranular stress corrosion cracking by polythionic acid in a stainless steel esterification column in an organic chemical plant, microstructure across weld and parent metal, (200X).
(Courtesy: Elayaperumal, in "Corrosion Failures in Process Industries in India: Statistical Analysis and Case Studies" in Special Supplement 2009, Corrosion Reviews Special Issue: India, England: Freund Publications, 2009, p 59, case study 10.)

SERVICE	Shell of a vertical column in which esterification takes place with alcohol as a reactant. Has been in operation for few years
PROBLEM	After one of the yearly maintenance operations while hydro testing heavy leakage occurred in several places of the shell near various weld joints
MATERIAL	316 L STAINLESS STEEL
OBSERVATIONS	Dye-penetrant test revealed multiple cracks on the inside surface on the weld regions both in the weld and parent metals. Metallographic test showed the cracks to be branched trans-granular
DIAGNOSIS	STRESS CORROSION CRACKING by POLYTHIONIC ACID which forms only during shutdown periods. Sulphur was a sudden impurity in the alcohol. It forms sulfide. The latter reacts with moisture and air during shutdown to produce PTA which induces SCC
REMEDY	Alkaline wash of equipments just before shutdown prior to exposure to air and moisture. Inert purging and/or blanketing for short shutdown periods

CASE STUDY 11   Half Pipe Limpet Coil of a Stirred Reactor in an Organic Chemical Plant

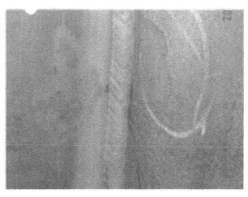

Case 11   Corrosion under insulation. Stress corrosion cracking of stainless steel half pipe limpet coil of a stirred reactor in an organic chemical plant.
(Courtesy: Elayaperumal, in "Corrosion Failures in Process Industries in India: Statistical Analysis and Case Studies" in Special Supplement 2009, Corrosion Reviews Special Issue: India, England: Freund Publications, 2009, p 60, case study 11.)

SERVICE	Limpet coil welded to the outside surface of a reactor used for both heating and cooling purposes
PROBLEM	Leaks occurred in the limpet coils within 1 year of service
MATERIAL	TYPE 304 STAINLESS STEEL
OBSERVATIONS	1. Dye-penetrant test showed multiple cracks on the coil near the weld joints 2. Metallography showed the cracks to be transgranular and branched 3. The mineral wool insulation covering the limpet coil was wet with high chlorides
DIAGNOSIS	EXTERNAL STRESS CORROSION CRACKING (ESCC) by chloride from a specific lot of re-used insulation
REMEDY	1. Always use fresh insulation whenever necessity comes for replacement 2. Use minimum heat input for welding the half pipe to the reactor 3. Consider duplex stainless steel as an alternative and better material of construction for the coil

---

CASE STUDY 12   Firewater Lines Buried Underground in an Organic Chemical Plant

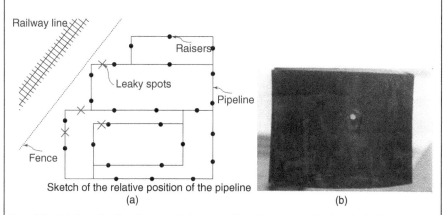

Sketch of the relative position of the pipeline
(a)   (b)

Case 12 (a) Sketch of underground firewater lines in an organic chemical plant showing positions (X marks) where leaks have occurred. (b) A leaky spot as seen on the outside surface of the underground fire water line pipe.

SERVICE	Firewater lines, 6″ NB, 4.8 mm WT, buried in underground soil, containing water inside under pressure, to be used whenever fire breaks out
PROBLEM	Leakage in isolated different places within a period of about 10 months
MATERIAL	CARBON STEEL
OBSERVATIONS	At the leaky areas, the outside surface had a large pit of size 15×25 mm, gradually reducing towards the inside surface resulting in a leaking hole
	The isolated pits were in one segment of the entire pipeline system, lying close to the backside fence of the plant, not in other segments
	Metallographical study indicated heavy intergranular corrosion
DIAGNOSIS	STRAY CURRENT CORROSION, due to induced leakage current from the electric traction railway line outside and very close to the fence
REMEDY	If it is not possible to plug the leakage current, weld scrap iron anodes at the leakage spots or raise the particular segment above ground

CASE STUDY 13   Alcohol Superheater in a PVC Manufacturing Petrochemical Plant

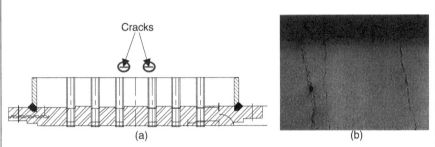

(a)   (b)

Case 13   (a) Longitudinal sketch of the bottom portion of tubes and bottom tube-sheet showing the location of the circumferential cracks in alcohol vapor super-heater. (b) Microstructure of the longitudinal cross section of the inside surface of a failed tube of the alcohol vapor super heater showing multiple cracks initiating on localized pits, (50X).

SERVICE	Vertical shell and tube heat-exchanger used for superheating alcohol vapor. The vapor enters at the tube-side bottom at 115°C and exits at the top at about 185°C. High pressure steam flows on the shell side
PROBLEM	Leakage occurred in about 30% of the tubes after only 7 months of operation
MATERIAL	TYPE 304 STAINLESS STEEL, seamless tubes, 19.05 mm OD, 2.77 mm WT and 2.5 m length
OBSERVATIONS	Circumferential cracks were present in all the leaky tubes at a height of about 5 mm from the top side of the bottom tube sheet. Localized deep pits and trans-granular branched cracks initiated at these pits on the inside surface and propagated towards outside surface to leakage. The input alcohol vapor contained liquid alcohol, moisture, and also chloride at about 50 ppm
DIAGNOSIS	STRESS CORROSION CRACKING by chloride present in the alcohol as impurity
REMEDY	Ensure that alcohol vaporizer upstream is fully efficient so that liquid carry-over to the superheater does not occur Also consider duplex stainless steel (UNS S31803) as an alternate MOC for the tubes

CASE STUDY 14   Package Boiler Tubes in an Alcohol Distillery Plant

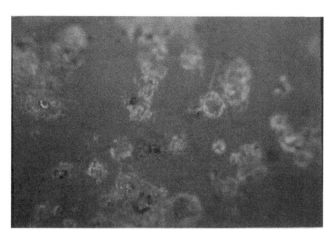

Case 14   Close-up view of the pits on the outside surface of one of the affected package boiler tubes.

SERVICE	Fire tube boiler in which fuel gas flows within the tube while the boiler feed-water along with condensate from the plant flows on the shell-side
PROBLEM	Heavy localized rust trees (tubercles) were seen on the outside surface (shell side surface) of several tubes after about 11 months of operation. Leakage followed at these places after additional 2 months of operation
MATERIAL	Boiler quality CARBON STEEL
OBSERVATIONS	Localized deep pitting corrosion on the outside surface. Pits contain loose brownish rust (corrosion product). These pits have grown in depth towards the inside surface and resulted in leakage. Rust contained chloride and was acidic in nature (pH of 1% solution was 3.5)
DIAGNOSIS	PITTING CORROSION due to ingress of chloride impurity to the boiler water through distillery process chemicals via the condensate
REMEDY	Identify and plug the source of leakage of chloride in the distillery process equipments

---

CASE STUDY 15   Reducers in a Reformer Tube in an Ammonia Plant of a Fertilizer Industry

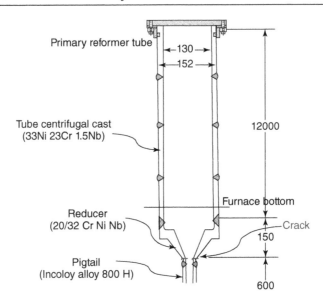

Case 15   Cross section of a reformer tube showing crack at the reducer at the bottom.

SERVICE	Reducers connecting primary reformer tubes to the downstream pig-tail tubes carrying reformed gas at a temperature of about 730°C
PROBLEM	Sudden leakage after about 10 years of service in 2% of the total number of tubes, through circumferential cracks in the reducer bottom
MATERIAL	ALLOY-800H as per ASTM B-408
OBSERVATIONS	• Cracks only at the smallest cross-section, just above the bottom weld connecting the outlet pig-tail to the reducer • Heavy carburization on the inside surface, resulting in brittle carbides • Failed reducers were made of castings with porosities while non-failed ones were of wrought sound forgings
DIAGNOSIS	CARBURIZATION along the porosities of the cast reducers forming brittle carbides which lead to cracking at the smallest cross-section
REMEDY	Replace cast reducers with sound defect-free forged ones

CASE STUDY 16   Pressure Safety Valve (PSV) Fitting on Instrumentation Tubes in an Off-shore Platform in an Oil and Gas Company

Case 16   Carbon steel pressure safety valve (PSV) fitting on stainless steel instrumentation tube in an off-shore platform in an oil and gas company.

SERVICE	PSV fitting on Instrumentation and other tubes in an off-shore platform
PROBLEM	The fittings showed very severe premature corrosion (rusting) while the tubes themselves did not show any corrosion sign
MATERIAL	316 STAINLESS STEEL for TUBES and CARBON STEEL forgings for fittings
OBSERVATIONS	• Very severe corrosion with collection of brownish rust throughout the fitting surface • Chemical analysis indicated that the fittings were made of carbon steel and not the stainless steel while the tubes were conforming to 316 SS
DIAGNOSIS	GALVANIC CORROSION of carbon steel in contact with stainless steel in a marine atmosphere
REMEDY	Ensure compliance with 316SS of both fittings and the tubings

CASE STUDY 17   Water Drum (Mud Drum) Shell in a Coal-Fired Steam Boiler

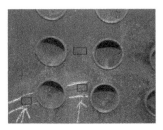

Case 17   Cracks on mud drum shell in a coal fired steam boiler.

SERVICE	Boiler produces steam at $56\,kg/cm^2$ pressure at a rate of 30 TPH. Out of the three boilers operating, one showed the problem of this case study
PROBLEM	Leakage of the shell of the water drum occurred after 7 years of operation. Further two leakages occurred within the next 1 year
MATERIAL	CARBON STEEL as per SA 302 Gr.B for drum and boiler quality carbon steel for tubes
OBSERVATIONS	• Cracks on the shell on the ligaments between adjacent tubes, easily revealed through dye-penetrant test on the inside surface (water side) and with difficulty in approach on the shell side (fuel side) • Intergranular corrosion initiated on the shell side (fuel side) and propagated towards the inside surface (water side) opened up as cracks • Visible corrosion on the boiler tube external surface at regions near the shell cracks
DIAGNOSIS	INTERGRANULAR CORROSION from the fuel side by POLYTHIONIC ACID which forms during shutdown periods only, as a result of high sulfur in the fuel. Ash accumulation in certain localized places initiates this attack
REMEDY	1. Caustic wash on the fuel side before opening up the boiler and exposure to atmospheric air and moisture during shutdown 2. Redesign for efficient soot blowing

CASE STUDY 18   Tubes in a Kettle Re-boiler of an Amine Plant

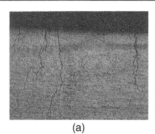 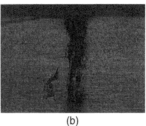

(a)   (b)

Case 18   (a) Caustic stress corrosion cracking of carbon steel reboiler tube in an amine plant (microstructure 50X). (b) Corrosion fatigue crack of carbon steel reboiler tubes in an amine plant (microstructure 50X).

SERVICE	U-tube kettle re-boiler. Various amine solutions are re-boiled on the shell side at temperatures ranging from 120 to 210°C with thermic fluid flowing on the tube side
PROBLEM	Tube leakage within 9 months of service in a replaced bundle and 18 months of service in the first bundle
MATERIAL	First bundle: 316 STAINLESS STEEL, tube size 25.4 mm OD × 2.03 mm WT Second bundle: 316L SS of same size
OBSERVATIONS	• Many tubes out of the total installed 15 tubes showed leakage • Circumferential cracks of the portions between tube-sheet and the first baffle. Cracks only on the top half portion of the tube bundle • Microscopic examination showed a few straight cracks and a few branched cracks. These were all multiple initiations on shell side and were transgranular in propagation
DIAGNOSIS	Straight cracks are CORROSION FATIGUE while the branched cracks are of STRESS CORROSION CRACKING. The causative chemical agent is CAUSTIC ALKALI present in the process medium. Stresses due to cantilever action and vibration during vigorous boiling
REMEDY	Replace with 316 SS with a thicker wall. 2.11 mm instead of 2.03 mm. 316 SS is stronger than 316L SS

CASE STUDY 19   Evaporator Tubes in an Organic Chemical Plant

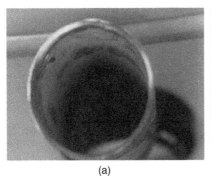 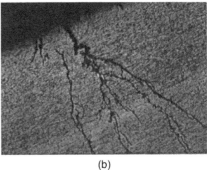

(a)   (b)

Case 19   (a) Calcium chloride evaporator tube showing cracks at one end. (b) Transgranular stress corrosion cracking from the inside surface (200X).

SERVICE	Calcium chloride evaporator tubes. Tube side: $CaCl_2$ solution, 52%, 150°C Shell side: steam, 7 bar, 170°C
PROBLEM	10% of tubes leaked within about 40 hours of trial operation
MATERIAL	Type 316L SS welded tubes (316 L/A 249) 38.1 mm OD × 1.65 mm WT
OBSERVATIONS	• Leak locations very close to either ends of the tubes within the length where the tubes were roll expanded into the tube sheet • Localized rust marks along the internal surface on the roll-expanded length • Leaks were through cracks initiated on the inside surface near the tube ends • Cracks were transgranular branched with multiple locations of initiation
DIAGNOSIS	(i) CHLORIDE STRESS CORROSION CRACKING (CSCC) from the inside surface by calcium chloride solution at the roller-expanded portions (ii) Rust marks from embedded iron from improper roll expanding tool
REMEDY	1. Subject new tubes to "Stress Relief Annealing" in addition to "Solution Annealing" as a procurement procedure before installing 2. Join the tubes to tube sheet through welding without roller expansion 3. Consider duplex stainless steel UNS S31803 for tube MOC as a long-term choice

CASE STUDY 20   Top Tube Sheet Vent Equalizer Weld Zone of a Gas Cooler in a Petrochemical Plant

GAS COOLER VENT EQUALIZER WELD JOINTS IN TOP TUBE SHEET

SERVICE	Four vent holes in the top tube sheet of a shell and tube gas cooler, connected to a steam drum. Hot gas in tube side and boiler feed water on shell side
PROBLEM	Within 1 month of operation, leakage of boiler feed water on external periphery of the top tube sheet at an area surrounding one of the vent nozzle weld joints
MATERIAL	Type 316 SS
OBSERVATIONS	• Cracks were found on the leaky areas. Cracks got initiated on a few places on the interior surface of the vent hole passage within the tube sheet and propagated in a branched manner through the ligament wall of the tube sheet • Chloride in the BFW was 20 ppm as against specified <1.00 ppm • Oxygen scavengers not regularly added • Insulation was absent in the affected areas leading to condensation
DIAGNOSIS	Chloride STRESS CORROSION CRACKING (SCC) due to improper operation of water treatment procedures
REMEDY	1. Maintain Cl^- level in BFW below 1 ppm 2. Ensure that oxygen scavenger is sufficiently added 3. Ensure presence of insulation wherever needed

CASE STUDY 21   Bottom Row Tubes in a Kettle Re-boiler of an Organic Chemical Plant

(a)

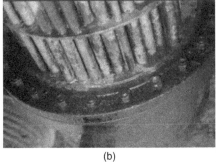
(b)

Case 21   (a) Bottom row tubes in an amine kettle re-boiler of an organic chemical plant. (b) Closer view of the external tubes with amine sludge.

SERVICE	Kettle reboiler with U tubes below a distillation column Shell side: aqueous solution of an amine. 60–120°C Tube side: thermic fluid
PROBLEM	Premature leakage within a month of operation from the bottom most rows of tubes with corrosion attack on rest of the bottom half of the tube buddle
MATERIAL	Carbon steel: seamless U tubes as per ASTM A-179: 25.4 mm OD × 2.11 mm thick
OBSERVATIONS	• The position of leakage on the tube length was close to the first baffle and to the tube sheet at the bottom near the thermic fluid entry side • Corrosive attack from the outside surface, amine side, in the form of deep grooves (gouging). Similar attack on the tube sheet bottom also from the amine side • Attack only on the lower half of the tubes and not on the upper half • Plenty of sediments and sludge at the bottom
DIAGNOSIS	CAUSTIC GOUGING attack by alkaline amine solution entrapped within the bottom sludge, during the process of bubbling upwards
REMEDY	Periodic thorough cleaning of the bottom sediment

CASE STUDY 22   Cages for Filter Bags in an Inorganic Chemical Plant

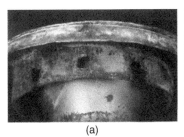 

(a)   (b)

Case 22   (a) Pitting and general corrosion in top flange of stainless steel filter bag hanger. (b) Cracking in cage rod weld joints.

SERVICE	Top ventury sheets and cage rods of bag filter cages of an inorganic chemicals plant producing silica powder. Temperatures vary from 120°C at the bottom inlet to 70°C at the top air/water vapor outlet
PROBLEM	Thinning down and collapsing of ventury sheets and cracking of cage rods, thereby spoiling the filter bags, all within a few months of operation
MATERIAL	316L SS
OBSERVATIONS	• Heavy thinning, localized pitting, and crevice corrosion on the upper funnel of the ventury • Cracks on bent surfaces of the ventury • Preferential corrosion at the weld joints • Cracking of cage rods near the weld joints • Feed is slightly acidic, pH ~5.5 with free residual chlorine 3–5 ppm • Materials of construction do not conform to 316L SS • No damage of any kind at the bottom half of the cage
DIAGNOSIS	• MOCs do not satisfy 316L SS, they are poorer in corrosion resistance than 316L SS • Improper welding of the cage-rod assembly. Instead of spot welding, welding with filler metals other than 316L SS has been adopted • Temperatures in the upper half are lower than dew point
REMEDY	1. Ensure quality of MOC 2. Adhere to total insulation even at the top 3. Add chloride-free biocide to process water

CASE STUDY 23   High Temperature Generator (HTG) Tubes of Vapor Absorption Chiller of an Air-Conditioning and Refrigeration Unit

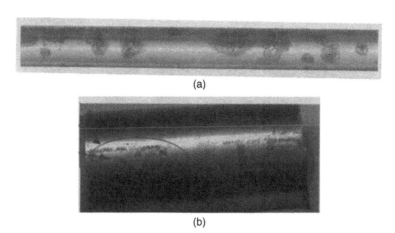

(a)

(b)

Case 23   (a) Inside surface of 430SS high temperature generator tube showing corrosion rust patches. (b) Outside surface of 430SS high temperature generator tube showing leaked areas.

SERVICE	Heat exchanger tubes for producing high temperatures inside a vapor absorption chiller machine
PROBLEM	Tubes showed leakage during pressure testing after installation and prior to usage
MATERIAL	Ferritic 430SS, welded construction as per ASTM A-268, 19.00 mm OD × 1.00 mm WT
OBSERVATIONS	• Very large percentage of tubes leaked • Corrosion initiated on the inside surface and propagated to outside surface resulting in leakage • Inside surface showed several localized rusty brown corrosion patches surrounding leaky spots with a shiny white etched background surface • Walls of corrosion trenches were very sharp and spiky • Calculated corrosion rate was very high, ~24 mm/year
DIAGNOSIS	Tubes have gone through an un-scheduled "acid cleaning/pickling" step which went un-recorded. The remaining acid has not been thoroughly washed and neutralized, resulting in localized corrosion patches
REMEDY	Avoid acid pickling/cleaning on finished tubes of 430SS

CASE STUDY 24   Gasket Seat in a Shell and Tube Condenser in a Petrochemical Plant

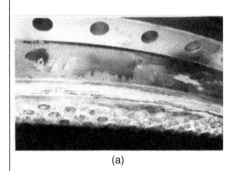
(a)

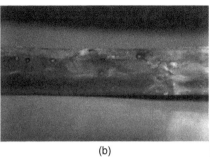
(b)

Case 24   (a) Gasket groove surface showing localized corrosion marks. (b) 304SS asbestos gasket envelope showing through and through corrosion pits.

SERVICE	Shell and tube assembly to condense organic product vapors flowing on the tube side, with cooling water, fresh land source water, on the shell side
PROBLEM	Leak has occurred on the cooling water side on the floating tube-sheet end. Gasket groove on the tube sheet face had corroded resulting in a rough surface and hence leak
MATERIAL	Type 304SS for tube-sheet and gasket
OBSERVATIONS	• Rough corrosion on the gasket seat groove • Gasket was specified as "Type 316 SS spiral wound PTFE filled." But the actual gasket used was "Type 304 SS asbestos enveloped" • The asbestos was in a highly damaged condition and the 304 SS envelope was in a highly pitted condition with several leaky holes
DIAGNOSIS	Pitting corrosion by the chlorides in the cooling water (~150 ppm) at about 40°C on the 304 SS envelope. This makes the asbestos wet. Wet asbestos with stagnant water leads to CREVICE CORROSION of both 304 SS envelope and 304 SS gasket groove
REMEDY	Use only corrosion-resistant nonabsorbing gaskets like "SS 316 Spiral Wound PTFE filled gasket" and not any lower quality absorbing type gaskets

CASE STUDY 25   Acid Gas CO$_2$ Cooler Condenser in Ammonia Plant of a Fertilizer Unit

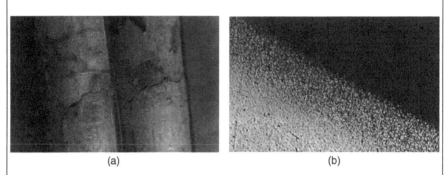

| (a) | (b) |

Case 25 (a) Outside surface of SS 304 cooler tubes showing circumferential cracks. (b) Optical micrograph of the longitudinal section of the tube showing carburization on the inside surface to a great depth (100X).

SERVICE	Shell and tube multi-pass cooler/condenser for cooling/condensing CO$_2$ gas and water vapor on the shell side by cooling water flowing on the tube side
PROBLEM	Leaking of tubes in the final pass
MATERIAL	304L SS
OBSERVATIONS	• Corrosion and cracks initiated on the tube-side, cooling water side • Tube wall skin temperature is maximum at the leaked portions: both on the shell side and on tube side • Heavy intergranular corrosion on the inside surface • Transgranular branched cracking initiated on the corroding front and propagating towards the outside surface • High carbon content on the inside surface. 0.18% as compared to specified 0.03% max.
DIAGNOSIS	Poor quality of the tubes: carbide precipitation on the inside surface during tube manufacturing: This results in INTERGRANULAR CRACKING (IGC) on the inside surface by cooling water, propagating to cracking by chloride of cooling water
REMEDY	Procure stainless steel tubes without the presence of "Carbide Network" throughout the wall thickness

CASE STUDY 26   Naphtha Coolers in a Fertilizer Plant

WIDE SPREAD PITTING CORROSION ON COOLING WATER SIDE

SERVICE	Shell and tube coolers. Naphtha with and without H_2S gas on the shell side and cooling water on the tube side
PROBLEM	Leakage
MATERIAL	CARBON STEEL, as per A-179, seamless
OBSERVATIONS	Localized corrosion from inside to outside. Shallow and wide pits on the inside. Coolers were used intermittently, not continuously, depending upon the need. Coolers are located at the farther end of the cooling water circuit, farthest from cooling tower and cooling water pumps. Velocity in these coolers was very low
DIAGNOSIS	"UNDER DEPOSIT PITTING CORROSION" on the cooling water side due to low velocity intermittent operation
REMEDY	Drain the water from these coolers prior to shutdown and maintain them in the most possible dry condition during the shutdown period

CASE STUDY 27   U Type Jet Dyeing Machine in a Textile Dyeing Unit

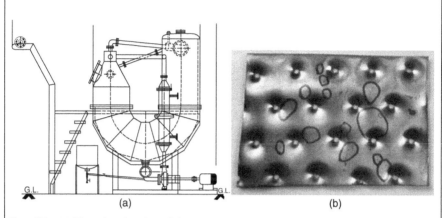

(a)   (b)

Case 27   (a) Elevation drawing of the U shape jet dyeing machine. (b) Localized sharp pitting corrosion marks on the inside surface of the perforated basket of jet dyeing machine.

SERVICE	Textile dyeing machine. U type construction. Perforated basket on the inside to contain the moving fabric in the dye solution
PROBLEM	Sharp pitting corrosion on the inside surface of the basket at the bottom half after 10 months of operation, thereby spoiling the fabrics being dyed in the machine
MATERIAL	Type 316L SS
OBSERVATIONS	Pitting corrosion only on the inside surface of the basket, not outside, facing the shell, that too on the bottom half portion of the machine, not on the top half. Edges of the pits were very sharp
DIAGNOSIS	CREVICE CORROSION caused by contact with wet fabric under stagnant conditions (without flow) when the machine was not in operation but with wet fabric lying inside. Wet fabric lies at the bottom of the basket touching only the inner surface of the basket
REMEDY	• Avoid long stagnancy period, with or without the fabric inside • Avoid contact with wet fabric during shutdown periods

CASE STUDY 28   Acetic Acid Manufacturing Unit in a Petrochemical Plant

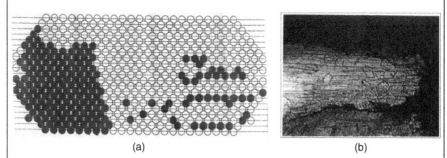

(a)   (b)

Case 28   (a) Cross section of the leaked tube bundle highlighting the positions of the leaked tubes. (b) Longitudinal microstructure of the tubes across the leaked portion showing intergranular corrosion (IGC), 200X.

SERVICE	Single pass horizontal heat exchanger exchanging heat from the hotter reactor product outlet gas mixture (Reactor Gas) on the tube side to the cooler recycle feed acetic acid gas (Recycle Gas) on the shell side in a counter current mode
PROBLEM	About 33% of the tubes leaked after about 4½ years of operation
MATERIAL	316 L SS. Seamless as per A-213. 19.05 mm OD × 1.65 mm W.T × 4600 mm long
OBSERVATIONS	• Most of the leaked tubes were in the lower half and on the right and left one-third of the tube bundle • Leaky pin-holes as seen on the outside surface of the tubes • All the leaky spots are roughly at one vertical plane, about 1 m from the reactor gas outlet end • Corrosion from inside to outside • Intergranular corrosion and carbide precipitation on the inside surface
DIAGNOSIS	CARBIDE PRECIPITATION during tube manufacturing resulting in INTERGRANULAR CORROSION on the inside surface at the cooler part of the tube bundle
REMEDY	The tubes must be free from carbide network both on the inside and outside surfaces before installation

CASE STUDY 29   Large Stainless Steel Pipeline in a Urea Plant of a Fertilizer Unit

 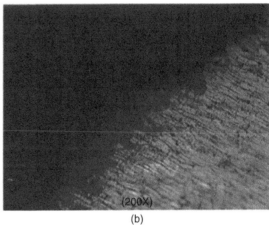

(a)   (b)

Case 29   (a) The burst 316L SS welded pipe on location. (b) Optical micrograph of the longitudinal cross section across the split surface of the pipe showing crevice corrosion (200X).

SERVICE	14″ NB 8.00 mm thick stainless steel pipeline transporting off-gas solution containing ammonia, carbon-dioxide, and ammonium carbamate at a pressure of 18 atm and at a temperature of 150°C
PROBLEM	Sudden bursting along the weld seam after 3 years of service
MATERIAL	316 L SS welded pipe as per ASTM A-358 Class-1
OBSERVATIONS	• Bursting along the interface between parent metal and the longitudinal weld seam • Lack of fusion at the root gap on the inside surface, as shown by radiographs and ultrasonic tests • Crevice corrosion on the process side at the root gap, progressing slowly towards the outside surface
DIAGNOSIS	QUALITY DEFICIENCY in terms of Lack of Fusion at the Root Gap on the inside surface followed by CREVICE CORROSION at the root gap over the years resulting in bursting at the operating high pressure
REMEDY	Procure welded pipes with full 100% radiography to ensure freedom from weld defects all along the weld seam and the interfaces

CASE STUDY 30   Heat Recovery System of a PVC Unit in a Petrochemical Plant

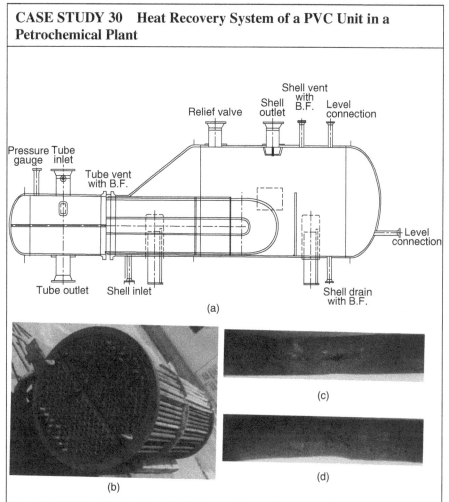

Case 30   (a) Elevation drawing of the kettle type heat exchanger. (b) Cross section of the tube bundle showing the plugged leaky tubes. (c) The outside surface of an affected tube showing corrosion patches. (d) The outside surface of a second affected tube showing corrosion patches.

continued

CASE STUDY 30	Continued
SERVICE	Kettle type heat exchanger with reactor outlet gas on the tube side, with inlet and outlet temperatures 360 and 160°C, respectively, and raw Alcohol from day tank on the shell side, with inlet and outlet temperatures 35 and 135°C, respectively
PROBLEM	Frequent failing through tube leakage within 6 months of service
MATERIAL	C.S, Seamless tubes as per A-179, U tubes, size 19.05 mm OD × 2.11 mm WT × 1410 mm straight leg length
OBSERVATIONS	• Only the first two rows of tubes facing the inlet alcohol on the shell side were affected at the tube-sheet end • The positions of leakage points were between the tube-sheet and the first baffle only, facing the alcohol entry nozzle • Patches of heavy deposits of different colors, black brown and white, with the leaky hole in the centre were seen on the outside surface (shell side surface) of the tubes. These were not seen on the inside surface • Deposits were acidic in nature and contained carbon, calcium, and magnesium
DIAGNOSIS	"UNDER DEPOSIT PITTING CORROSION" initiated on the outside surface of the tubes. Carbon deposit due to intense burning of alcohol at the localized entry point and calcium and magnesium deposits from the impurity water present in the raw alcohol
REMEDY	Periodic cleaning of the outside surface as frequently as possible, not less than once in a month

CASE STUDY 31   EDC Furnace Coil of a PVC Plant in a Petrochemical Unit

Stack temp: 290°C

40'C\30 kg/cm^2

520'C

20 kg/cm2

Quench tower

Convection bottom 410'C

Furnace feed tank

F.F. Pump

Kerosene burner

Fire box temp 710'c

Note: 8 passes removed from convection section

Convection: 18NO.S. finned tubes

Shock section: 3 plain tubes

Radient section: 27 plain tubes

(a)

(b)

(c)

Case 31   (a) Flow diagram for EDC heating. (b) Corrosion of fins on the outside surface of EDC furnace coil tubes. (c) Stress corrosion cracking initiating on the outside surface of EDC furnace coil tubes, optical micrograph, (100X).

continued

CASE STUDY 31	Continued
SERVICE	Convection section tubes of an EDC furnace coil. Outside fins. Ambient temperature EDC enters the tube inside at the top turn. Furnace flue gas on the shell side
PROBLEM	Top few coils leaked after 3 years of installation
MATERIAL	Type 347 SS for tubes and type 430 SS for outside fins
OBSERVATIONS	• Fins are affected by uniform corrosion resulting in wall thinning and breaking away • Multiple cracks on the outside surface of the tubes initiated at localized pits • Acidic Deposits on fins and tubes on the outside surface (pH≈2.00) • The fuel used in the furnace was kerosene with high content of moisture, sulfur, and chloride
DIAGNOSIS	• Heavy condensate corrosion on the outside fins of the tubes lying on the cold portions of the coil with temperatures less than the dew-point • Salts from the kerosene fuel deposit on the outside surface of the tubes and lead to chloride stress corrosion cracking
REMEDY	• Avoid condensation by modifying the temperatures in the affected regions to be above dew-point • Change the fuel to a better grade with lower inorganic impurities including moisture

CASE STUDY 32   Internals of a Stirred Reactor Processing Ortho Phosphoric Acid

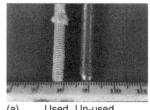 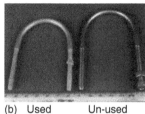

(a)   Used Un-used   (b) Used   Un-used

Case 32   (a) Close view of U-bent arm of 316L SS, both used, on left, and unused, on right. (b) Overall view of used and unused U-bends.

SERVICE	Stirred reactor with internal coil for heating and cooling. Processing batch reactions involving organic amines and orthophosphoric acid (OPA) at about 220°C
PROBLEM	• Thinning down of U-bolts on the threaded ends thereby loosening and losing the nuts within 18 months of operation • Etching of shell, inner coil, and agitator blades
MATERIAL	Type 316LSS
OBSERVATIONS	• Threads of the U-bolts thinned down with shining appearance with corrosion rate $\sim$1.00 mm/year • Plain portions thinned down with shining appearance with corrosion rate $\sim$0.30 mm/year • Bottom dished end corroded at a rate of about 0.57 mm/year • Agitator blades: 0.19 and 0.14 mm/year corrosion at bottom and top, respectively • Interior coil: lower turns 0.30 mm/year, Upper turns 0.10 mm/year corrosion rates • Shell wall: 0.05 mm/year
DIAGNOSIS	• UNIFORM CORROSION by OPA • Acceptable rate on the shell wall • Somewhat high rates on the agitator blades, bottom dished end, and interior coils with addition to dissolution rate by slurry erosion and/or temperature • Highest unacceptable rate on the bolt threaded portions due to cold work strains
REMEDY	• Avoid threaded nut-bolt construction for clamping bolts. Adopt a rigid welded construction using 316L SS filler metal • Monitor corrosion rates periodically and take advance action for timely replacements • Reduce temperature and RPM if process permits

CASE STUDY 33   Salt Evaporator

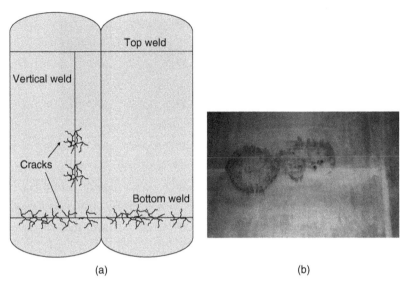

(a)   (b)

Case 33   (a) Sketch of the inside surface of the evaporator showing location of stress corrosion cracking in and around vertical and horizontal weld joints. (b) Photograph of a typical cluster of stress corrosion cracks around weld line on the inside surface.

SERVICE	Evaporation of a sodium chloride salt solution in a jacketed vessel to produce very fine powder of the salt: 114°C, pH≈5.6, and salt concentration 300gpl
PROBLEM	Cracks near the weld joints in a very short time
MATERIAL	Type 316 and 316L SS
OBSERVATIONS	Multiple branched cracks on the inside surface of the shell at the following locations: • Across the horizontal bottom weld line, (shell to bottom dished end weld line) • Across one single vertical weld line • At places corresponding to the welds on the outside surface (heat tint locations)
DIAGNOSIS	CHLORIDE STRESS CORROSION CRACKING
REMEDY	• Reduce residual stresses by using low heat input welding process • As a long-term basis, consider using duplex SS UNS S31803 as the material of construction for the evaporator

CASE STUDY 34 Cooler/Condenser Tubes of an Absorption Chiller Machine of an Airconditioning Plant

(a)

The galvanic series
Reactivity ranking of metals / alloys in seawater

More inert (cathodic) ↑
Platinum
Gold
Graphite
Titanium
Silver
316 stainless steel (passive)
Nickel (passive)
Copper
Nickel (active)
Tin
Lead
316 stainless steel (active)
Iron / Steel
Aluminium alloys
Cadmium
More active (anodic) ↓
Zinc
Magnesium

(b)

Case 34 (a) Erosion corrosion in cupro-nickel condenser tube in cooling water service. (b) Galvanic series in sea water, highlighting copper and graphite.

continued

CASE STUDY 34   Continued

SERVICE	Cooling tubes with cooling water on the inside and absorption medium on the outside
PROBLEM	Leakage through pinholes as seen on the outside surface
MATERIAL	90/10/02 cupro-nickel with iron
OBSERVATIONS	• Localized corrosion at several places on the inside surface, the cooling water side, very deep at locations where leaks have occurred • Erosion corrosion attack on the inside surface at leakage points near the cooling water inlet end • Chemical analysis of the tube material did not conform to tube specification ASTM B-111/UNS C70600. Iron content was only 0.44% as against the required 1.0–1.8% • Carbon, an element which is not expected, was found to present, that too at a high level of about 23%
DIAGNOSIS	• EROSION CORROSION at the inlet end due to insufficient iron • GALVANIC CORROSION of copper in contact with carbon, the major impurity present, at several places leading to localized corrosion of copper and leakage
REMEDY	• Ensure compliance to quality while purchasing the tubes • Impose carbon analysis of the tubes while purchasing

CASE STUDY 35   Nitro Mass Cooler in an Organic Chemical Plant

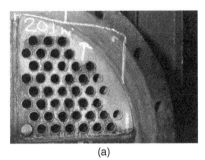 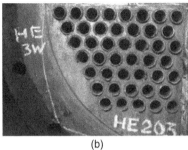

(a)   (b)

Case 35   (a) Inlet end erosion corrosion of carbon steel tubes in an acid cooler, first pass. (b) Inlet end erosion corrosion of carbon steel tubes in an acid cooler, second pass.

SERVICE	Shell and tube coolers, three in number, in series. Tube side: 60% H_2SO_4, 34% HNO_3, and 6% H_2O. 4 Passes on tube side. Shell side: calcium chloride brine at 0–2°C. The nitro-mass cools from 18 to 6°C
PROBLEM	Routine Inspection during the 14th year of operation showed severe corrosion in the first pass inlet ends of the tubes in each cooler
MATERIAL	Carbon steel. Seamless tubes as per specification A-179. Size: 25.4 mm OD × 10 SWG WT × 3.0 m long. 3.00 mm protrusion beyond the tube-sheet
OBSERVATIONS	• All the inlet portions of the first passes of each cooler were affected severely, eating away the 3 mm tube protrusions on the tube-sheet • Etching characteristics, indicative of erosive acid attack, on the entry side tube sheet surface • 50% increase in the production rate in the 13th and 14th year of operation, with corresponding increase in the mass flow rate and linear flow rate
DIAGNOSIS	EROSION CORROSION by concentrated acid at tube inlet portions
REMEDY	• Carry out a cost-comparison analysis and reduce the flow rate if possible • Grind-off the 3 mm protrusions to provide smooth ends with reduced turbulence at the inlet ends

CASE STUDY 36   Fertilizer Industry Ammonia Plant Natural Gas Feed Preheater Coil

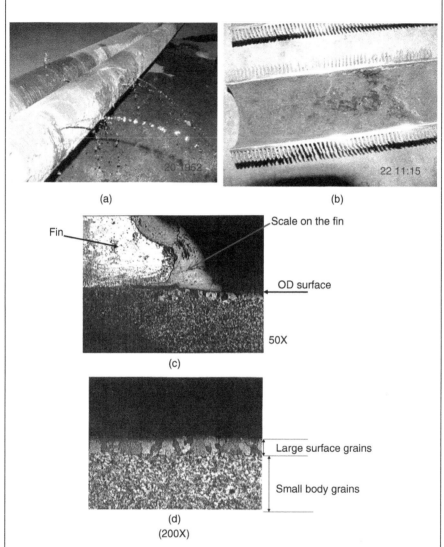

Case 36   (a) Pin-hole leakage during water testing of leaked alloy steel pipe. (b) Corrosion on the inside-surface of alloy steel pipe leading to pin-hole leakage, corrosion of external fins can also be noticed. (c) Longitudinal metallographic cross section across the pipe wall at the affected area including the fins (50X). (d) Enlarged magnification of the outside surface of the pipe wall showing large surface grains (200X).

CASE STUDY 36 Continued	
SERVICE	Coil preheats the feedstock Natural Gas (NG). The latter flows inside the pipe of the coil and gets heated from 30 to 371°C, while flue gas from the primary reformer flows around the outside surface of the pipes, giving away the heat to the pipes thereby the temperature gets reduced from 425 to 350°C
PROBLEM	Leakage through the 6 mm thick pipe in 2 years
MATERIAL	Alloy Steel P-11. Seamless pipe as per specification A-335. Size 4″ Sch.40 (4.5″ OD × 6.03 mm WT). Finned on the outside with carbon steel. Sixty-four pipes in all in the coil
OBSERVATIONS	• Pinhole leakage as seen on the outside surface • Heavy scaling followed by leaky hole formation on the inside surface • Heavy warping and scaling of the OD fins • The inside scale consisted of mostly the highly magnetic iron sulfide • Heavy grain growth on the OD surface • Grain boundary penetration of sulfur from the inside surface • The actual operating temperatures were much higher than the design temperatures. Flue gas in at 521°C, Flue gas out at 428°C, and NG out at 500°C
DIAGNOSIS	Excessive sulfidation on the inside surface due to both undetected high sulfur in the NG feed and operating temperatures much higher than the design temperatures
REMEDY	• Minimize sulfur in the feed gas • Control the operating temperatures within the design limits • Change the material of construction of the pipe of the coil 9Cr-1Mo alloy steel if the above controls are not possible

CASE STUDY 37   Petrochemical Unit. PVC Plant. Radiant Coils of the EDC Pyrolysis Furnace

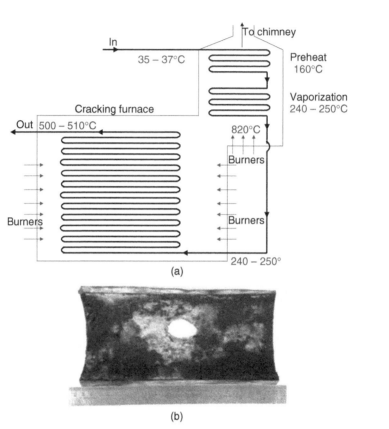

(a)

(b)

Case 37   (a) Sketch of the flow diagram of EDC through preheater, vaporizer and furnace. (b) Photograph of the inside surface of the radiant coil at a leaked area.

CASE STUDY 37 Continued	
SERVICE	Radiant coils of the furnace to heat EDC for Pyrolysis. EDC vapor at 250°C
PROBLEM	First few radiant coils leaked through the 10.60 mm thick walls in about 10 months of service
MATERIAL	Incoloy-800 HT (UNS N08811). Seamless pipe as per specification ASTM B-507. Size 6.0″ OD × 10.6 mm WT
OBSERVATIONS	• Heavy coke deposition, brownish corrosion product, and leaky hole as seen on the inside surface • Carbide formation on the inside surface • Calculated corrosion rate: 12.58 mm/year • Large quantities of free HCl acid found inside
DIAGNOSIS	Severe corrosion on the inside surface by concentrated aqueous hydrochloric acid. Moisture continues to be present in the EDC feed to the furnace due to nonfunctioning of the vaporizer upstream. The cracked EDC forms hydrochloric acid with the moisture present. Ferric chloride, formed due to HCl attack on the base iron of the pipe, accelerates the attack by HCl. Porosity in the coke lump encourages retention of HCl acid thus accelerating the attack
REMEDY	Avoid carryover of liquid EDC along with moisture to the pyrolysis furnace, by ensuring efficient operation of preheater and vaporizer

CASE STUDY 38   Exhaust Gas Boiler in a Sugar Mill

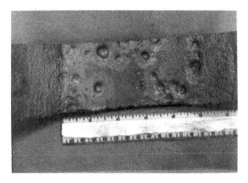

Case 38   Pitting corrosion due to dissolved oxygen in carbon steel boiler tube.

SERVICE	Waste heat boiler. Fuel: exhaust gas from burning of bagasse of a sugar mill along the outside surface of the tubes
PROBLEM	Leakage in the first, lower banks of convection tubes. No leakage in the second upper banks
MATERIAL	Boiler quality carbon steel
OBSERVATIONS	• General rusting, small size corrosion pits throughout and deep and wide pits in localized places on the inside surface • Outside surface free from corrosion • DM water without effective deaeration (either thermal or chemical) is used as boiler feed water. The dissolved oxygen content was 9–10 ppm as against the specified maximum of 0.005 ppm • The chemical oxygen scavenger sodium sulfite was only about 8 ppm as against required 71 ppm to scavenge 9 ppm of dissolved oxygen
DIAGNOSIS	CORROSION PITTING BY DISSOLVED OXYGEN in the feed water. Thermal deaeration was not practiced. Chemical deaeration was not sufficient and effective
REMEDY	• Add catalyzed sodium sulfite in sufficient quantities such that 10–15 ppm residual remains in the boiler blow-down water • Consider replacing sodium sulfite with catalyzed hydrazine

CASE STUDY 39   Domestic Storage Water Heater

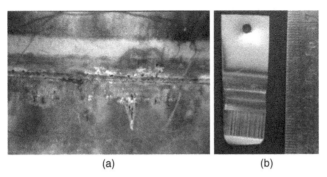

(a)   (b)

Case 39   (a) Corrosion on the inside surface of a stainless steel domestic storage water heater in and around weld joint. (b) Corrosion at and near the weld in a test coupon of stainless steel.

SERVICE	Domestic storage water heater
PROBLEM	Premature corrosion and leakage in the heater shells experienced by the end users
MATERIAL	304, 304L, 316, and 316L SS
OBSERVATIONS	• Pitting corrosion initiated on the internal surface near about the weld seams only • Pits have grown from inside to outside resulting in leaks of the stored water • Pits contained brownish rust inside them • Rust-laden brownish water trickled down with gravity giving a line of vertical pits. Similar appearance on the outside surface also • Pits initiated on the heat-tinted areas on either side of the weld • Attack was more frequent and prominent with hard waters than with soft waters • E_{pp} (pitting potential) was less noble for welded coupon than for plain coupon • Features were similar in all grades of SS used
DIAGNOSIS	Preferential corrosion on the heat-tinted oxide layer on either side of the welds
REMEDY	• Minimize heat tint by having inert atmosphere during welding • Carry out pickling process in a mixture of hydrofluoric and nitric acids followed by passivation in pure nitric acid solution

CASE STUDY 40   Stirred Reactor in a Rubber Chemical Plant

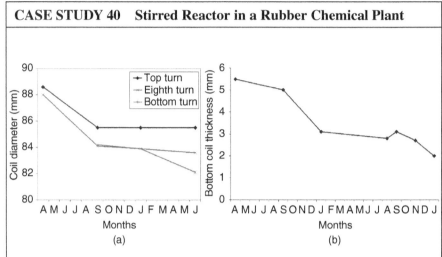

Case 40   (a) Decrease in coil diameter with time. (b) Decrease in coil thickness with time.

SERVICE	Batch reactor with internal agitator and internal heating and cooling coil. Processing PSA slurry as catalyst in organic solvents
PROBLEM	Leak in the bottom most coil
MATERIAL	316L SS
OBSERVATIONS	• Thickness reduction on the coil is higher and faster than that on the shell and bottom dished end • Diameter reduction on the top row of the coils is less than that at the bottom most coil • Thickness reduction rate at the bottom most coil was 1.99 mm/year, very high rate of corrosion of 316L SS in any chemical medium • Coil outside surface showed sharp cutting features
DIAGNOSIS	Fine particle, PSA, slurry erosion, more at the bottom than at the top because of agitator blades motion at the bottom. 316L SS is not resistant for solid particle slurry erosion
REMEDY	Use either 316Ti SS or Duplex SS (UNS S31803) in the place of 316L SS as the material of construction for the coil

CASE STUDY 41   Stainless Steel Tubes During Long Storage in Packed Condition

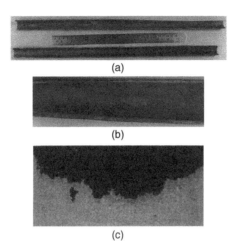

(a)

(b)

(c)

Case 41   (a) Overall view of the in-side surface of the affected tubes. (b) Close view of inside surface of an affected tube. (c) Longitudinal microstructure across the wall of an affected tube near the inside surface (200X).

SERVICE	Stainless steel tubes of size 25. mm OD × 2.11 mm WT × 9150 mm L, 480 tubes, kept stored in unused condition for about five and a half years
PROBLEM	Leaks during hydro-testing just prior to usage
MATERIAL	Type 405 Ferritic SS as per ASTM A-268
OBSERVATIONS	Overall rusted brown surface on the inside with localized intense corrosion patches leading to leaking pits.
DIAGNOSIS	Uniform Corrosion on the inside surface due to excessive moisture left behind as a result of incomplete drying prior to packing and due to the absence of moisture absorbing gel inside the packs. Type 405 ferritic SS is less corrosion-resistant than 300 series austenitics. It corrodes in atmospheres where moisture is present and oxygen is in restricted supply, such as the atmosphere on the inside surface of end closed and packed tubes where air circulation is absent
REMEDY	Do not store Ferritic SS tubes for long times in atmospheres where there is no supply of oxygen or atmospheric air

CASE STUDY 42   Fertilizer Plant. Ammonia Units. Secondary Waste Heat Boiler Tubes

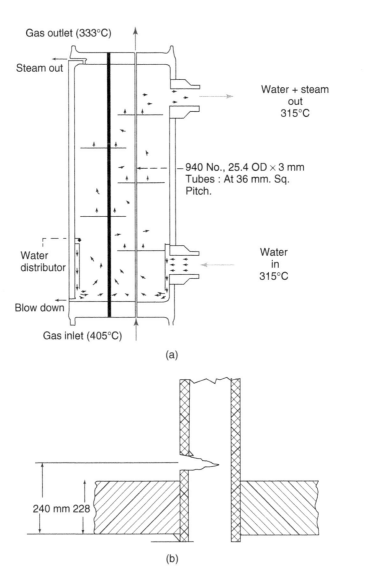

Case 42   (a) Flow pattern of process gas and boiler feed water in the waste heat boiler. (b) Approximate location of the corrosive attack on the outside surface of the waste heat boiler tube at the bottom portion.

CASE STUDY 42 Continued	
SERVICE	Vertical shell and tube waste heat boiler. Tube side: gas. Shell side: boiler feed water
PROBLEM	Leakage at the bottom of the tubes after $4\frac{1}{2}$ years of service in two units of identical design and operation
MATERIAL	Alloy steel
OBSERVATIONS	• Both the units showed identical problem • Corrosive attack at the bottom portion of the tubes, at a height of about 12–20 mm from the top surface of the bottom tube-sheet • Corrosion from the OD surface, shell side, feed water side, of the tubes proceeding towards inside surface, at positions little lower than the boiler feed water entry point at the bottom, • The corrosive attack was typical of caustic gouging, attack by caustic alkali, rust-free • Insufficient blow down from the very bottom areas of the shell side. Ineffective drain nozzles
DIAGNOSIS	CAUSTIC GOUGING corrosion at the bottom outside surface of the tubes due to improper drainage of the alkaline sludge formed at the bottom due to steam formation
REMEDY	Modify the design at the bottom in order to provide effective uniform drainage of the boiler blow down water from the shell side at the bottom

CASE STUDY 43   Hospital Hydroclave for Treating Wastes

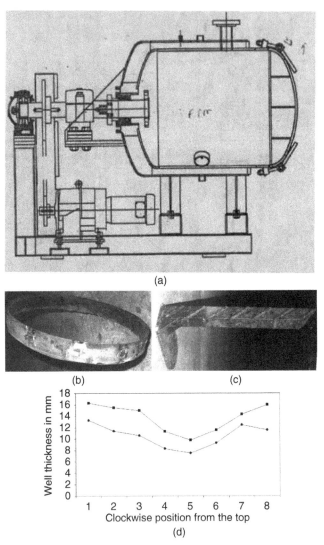

Case 43   (a) Elevation drawing of the hospital hydroclave for treating wastes. (b) Removed shell ring – heavy rusting on the inside surface. (c) Agitator arm after complete corrosion of the blades. (d) "Variation of shell thickness clock-wise from 1 O'clock to 8 O'clock positions."

CASE STUDY 43 Continued	
SERVICE	The wastes generated in hospitals are steam-sterilized, shredded, and discharged in the hydroclave. Temperature 121°C, pressure 15 kg/cm^2, and duration per batch 75 minutes
PROBLEM	Within 2 years of operation, heavy wall thickness reduction noticed at the lower half
MATERIAL	Carbon steel
OBSERVATIONS	• Pitting corrosion in O ring grooves and cover plates • Heavy thickness reduction starting from the inside surface. Uniform attack at the bottom half of the shell. Estimated corrosion rate: 4.8 mm/year • Rusting at the agitator arm, blades and blade joints and complete detachment of the blades • pH of the wetness of the processed waste after neutralization with sufficient caustic was around 6.5 indicating that the waste processed in the hydroclave is very acidic
DIAGNOSIS	UNIFORM CORROSION by the acidic wastes charged into the hydroclave without caustic neutralization
REMEDY	Add sufficient alkali to the waste mass before processing so as to bring the pH of the in-coming waste to about 10

CASE STUDY 44   Petrochemical Plant. Phosgene Absorption Column Internals

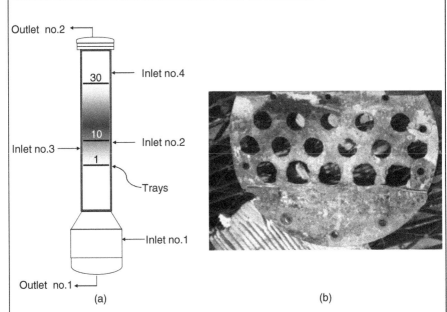

Outlet no.2

30   Inlet no.4

10   Inlet no.2

Inlet no.3

1

Trays

Inlet no.1

Outlet no.1

(a)   (b)

Case 44   (a) Schematic of the absorption column indicating the corroded region in the middle. (b) A typical corroded tray.

SERVICE	Valve tray contact between gas mixture (phosgene + HCl) entering at the bottom with organic solvent coming from the top. Pressure 1.7 kg/cm^{2}g, Temperatures: −8.0°C at the bottom and 35–40°C at the top
PROBLEM	After about 2 years of operation, heavy corrosion was observed on trays and valves at intermediate levels, with all valves missing and the trays heavily thinned down. Even the newly replaced trays and valves were found to be appreciably corroded within 1½ months of service

CASE STUDY 44	**Continued**
MATERIAL	Shell: carbon steel Trays: alloy steel 4360 Valves: 410 SS
OBSERVATIONS	• Heavy corrosion of middle trays and valves, numbers 6–18 • Bottom trays 1–5 and top trays 19–30 are not at all affected by corrosion • Bottom sludge contained mainly iron and iron chloride with appreciable presence of chromium, nickel, and cobalt • pH of the water extract of the sludge was about 2.5, highly acidic • Carbon steel of the shell in the affected elevation has corroded at a rate of 1.93 mm/year, a highly unacceptable rate
DIAGNOSIS	• Corrosion by an aqueous solution of hydrochloric acid (formed as a result of moisture present in the feed gas) as opposed to dry HCl gas, at temperatures 0°C and above. Top trays see dry HCl gas, so no corrosion, and the bottom trays are at too low a temperature for appreciable corrosion to occur • Galvanic contact between alloy steel trays and stainless steel valves accelerates the corrosion of the former
REMEDY	• Look into process changes to avoid the presence of moisture in the feed gas • Moist HCl being highly corrosive, change of MOC to one which completely resists moist HCl corrosion would not be cost-effective • Coating/lining and or replacement with Teflon or any other plastic material should be checked for their compatibility with the organic solvent and the gas used

CASE STUDY 45   Fertilizer Unit, Ammonia Plant, Start-Up Preheater Outlet Line

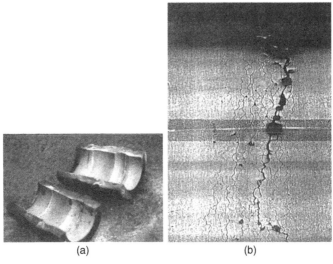

(a)   (b)

Case 45   (a) Dye penetrant test indications on the heat affected zones of the weld. (b) Intergranular corrosion and cracking initiated on the outside surface in the weld heat affected zone (200X).

SERVICE	Outlet pipes of start-up preheater of ammonia plant. Inside: syn-gas. Outside: atmospheric insulation
PROBLEM	Sudden cracking near the first weld joint with a reducer after about 18 years of service
MATERIAL	304H SS. 6 and 5″ NB pipes
OBSERVATIONS	• Crack near the weld joint on the heat-affected zone • Intergranular crack initiated on the outside surface
DIAGNOSIS	INTERGRANULAR CORROSION UNDER INSULATION (CUI-IGC). Sensitization while erection welding at a pre-commissioning stage followed by slow intergranular corrosion whenever the outside insulation gets wet. The minerals of the insulation get leached out into the moisture, form a corrosive medium, and corrode the sensitized grain boundary areas
REMEDY	• Use 316L SS as the MOC for the pipes • Ensure that insulation is always dry by properly designed protective cladding coverage around it

CASE STUDY 46 Petrochemical Plant, Underground Fire–Water Pipelines

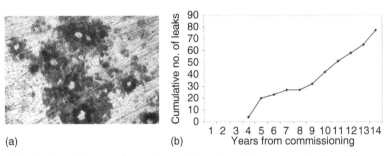

(a)   (b)

Case 46   (a) Microbial corrosion on the inside surface of fire water pipeline. (b) Increase of number of leaks with time in fire water pipe line.

SERVICE	Fire water pipelines. Partly submerged in soil underground and partly above-ground
PROBLEM	Leakage from third year of installation. Sudden and frequent leakages at locations near the pump house.
MATERIAL	Carbon steel
OBSERVATIONS	• Leaks in 12 o'clock and 3 o'clock positions • 70% of the leaks were at locations near the fire water pump house • Water remains stagnant in the pipes at a static pressure of 10 kg/cm^2 • Pipe outside surface exposed to the underground soil is under cathodic protection • All the leaks due to corrosive attack on the inside surface, water side • Localized corrosion in the form of pinholes with partly adherent loose scales surrounding the pinholes • Scale samples showed 17% L.O.I on dry basis. High amount of anaerobic micro-organism in the scales
DIAGNOSIS	• Micro-biological corrosion on the inside surface. Rate of corrosion increases with time • Leaks at locations near the fire water pump house, whenever the pump is started
REMEDY	• Keep the line water inside under the mode of periodic circulation instead of static mode • If shutdown is permitted, drain and blow with hot air to kill the bacteria before refilling

CASE STUDY 47   Monel Clad Evaporator in a Pure Water Plant

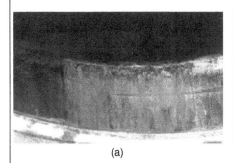
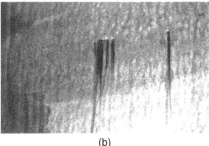

(a)   (b)

Case 47   (a) Preferential corrosion on the weld line in a Monel clad carbon steel vessel. (b) Corrosion on isolated points on shell course weld in Monel clad carbon steel vessel.

SERVICE	Salt slurry evaporator to produce pure water at 61, 84, and 105°C design temperatures and at 0.86 bar pressure
PROBLEM	Rusting and leaking on the Monel clad steel shell within 15 days of operation
MATERIAL	14 mm thick carbon steel with 2.5 mm thick Monel 400 clad vessel
OBSERVATIONS	• Butt welds are corroded with brownish rust oozing out • No corrosion in the Monel-to-Monel welds • Iron content in the welds was very high: 55.6% as against the allowable 2.5%
DIAGNOSIS	• Localized corrosion in the clad welds due to excess iron present in the weld metal
REMEDY	• Clad with transition metal • A barrier layer of nickel filler metal "61" (AWS-ERNi-1) should be applied first followed by Monel filler metal "60"

CASE STUDY 48 Loop Steamer Machine in a Textile Dyeing Unit		

 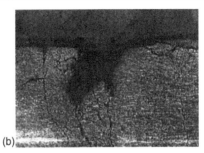

(a)   (b)

Case 48   (a) Stainless steel inside surface of the loop steamer machine. (b) Pitting corrosion and stress corrosion cracking by chloride initiated on the inside surface of stainless steel chamber.

SERVICE	Textile cloths are processed in the machine through rolls for dyeing, printing, and steaming (at 150–180°C)
PROBLEM	Premature corrosion (general corrosion creating roughness, pitting and cracking) in process wetted parts
MATERIAL	Type 316L SS
OBSERVATIONS	• General uniform corrosion throughout the wetted surfaces, creating grayish black corrosion products/deposits, to the extent of choking of spaces of limited widths • Deep corrosion pits below the deposits. Pits were not present on surfaces free of deposits • No corrosion upstream of the steamer machine. Within the machine, corrosion has taken place only in places where steaming occurs • Uniform and localized pitting corrosion and chloride stress-corrosion-cracking below the pits • Impure painting/dyeing chemicals: Citric acid pH 1.2 as against 1.9 in the pure acid, tartaric acid pH 1.1 as against 1.8 and high chloride level, up to 3.9% and considerable amount of sulfates
DIAGNOSIS	• Harmful chemicals present in high proportions in the in-coming dyeing/printing media lead to different corrosion phenomena
REMEDY	• Adopt periodic cleaning of the surfaces so that deposits (due to action of steam on the dyeing/printing media) are not allowed to accumulate on the wetted surfaces • Check for harmful impurities in the chemicals being added to the process

CASE STUDY 49   Rubber Chemicals Plant: Leakage in Process Pipelines

GENERAL CORROSION OF CARBON STEEL PIPELINE ON HORIZONTAL SECTIONS BY ORGANIC ACID

SERVICE	Outlet piping from a separator to autoclave processing certain organic reactions
PROBLEM	Leakage on the pipeline within 25 days
MATERIAL	Carbon steel
OBSERVATIONS	• Noticeable general uniform corrosion attack on the separator and autoclave by hot H_2S gas • On the outlet pipeline, only horizontal portions show major corrosion resulting in leakage. No noticeable corrosion on the vertical portions
DIAGNOSIS	• Uniform corrosion by an organic acid (sulfur-related carbamic acid), severe when it remains stagnant or where there is restricted flow
REMEDY	• Review process details and modify to avoid un-wanted acid formation • Replace the pipelines with 316L SS • Periodically monitor the thickness of both the SS pipelines and the carbon steel vessels

CASE STUDY 50   Clay Drier in a Clay manufacturing Plant

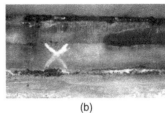

(a)   (b)

Case 50   (a) The outer shell of the clay drier. (b) Typical corroded areas inside the drier.

SERVICE	Drying of wet clay in a rotary drier, temperature varying from 100 to 600°C, by a partly co-current flue gas entering at 1100°C and returning at 200°C
PROBLEM	Heavy corrosion damage near the wet clay entry portions, about 1–2 m out of the total length of about 8–9 m
MATERIAL	Carbon steel, later changed to 304 SS
OBSERVATIONS	• Severe general uniform corrosive attack on carbon steel on portions near the wet clay entry, on locations contacted by the wet clay. Brown rust patches with heavy loss of thickness • Aluminium spraying on the corroded areas has not given any improvement in corrosion resistance • Replacement of the affected areas with 304 SS has resulted in stress corrosion cracking • The wetness associated with the clay gave the following analysis: pH 2.3, TDS 1500 ppm, Chloride 250–500 ppm, and TH 400 ppm • No evidence of any corrosion on the rest of the portions of the drier in the balance 6–7 m on the high temperature zones
DIAGNOSIS	• GENERAL UNIFORM CORROSION by the highly acidic water of the wet clay before the associated moisture evaporates into steam • The chlorides in the wetness of the clay have resulted in stress corrosion cracking of the replaced SS portions
REMEDY	• Change the MOC in the affected length of the drier to Duplex SS UNS S31803

CASE STUDY 51   Hydrogen Sulfide Processing Plant

(a)   (b)

(c)

Case 51   (a) Corrosive attack as seen closely on the outside surface of three of the affected tubes. (b) Corrosive attack as seen on the outside surfaces of several tubes. (c) Sub-surface blister bursts close to the outer surface of the tubes, optical micrograph (50X).

SERVICE	Horizontal shell and tube heat exchanger. Process water saturated with H_2S, flowing on the tube side, gets heated from 114 to 170°C at a pressure of 20 kg/cm^2 by steam flowing on the shell side at a pressure of 16 kg/cm^2. Steam gets condensed and flows out of shell side
PROBLEM	After 8 years of service, several tubes on the top half showed bending and severe localized pitting on the outside (shell side) surface. Crack-like features were seen emanating from several of these pits. Many tubes, without showing any visible attack, got rejected during eddy-current testing
MATERIAL	316L SS seamless tubes as per ASTM A-213 specification, 20 mm OD × 1.65 mm WT × 4900 mm L × 774 Nos.

CASE STUDY 51	**Continued**
OBSERVATIONS	• Affected tubes are from the top half of the bundle and showed the corrosion pits and cracks on the outside surface (shell side) • The affected areas are between the floating tube-sheet end and the first baffle approximately near the vertical plane corresponding to the shell side condensate outlet • Feed water upstream for producing steam at the Intermediate Steam Generator (ISG) contains process condensate also along with DM water make-up
	• Condensate from ISG showed an acidic pH of 3.5 as against the set range of 7.0–9.5 and presence of sulfide (70–935 ppb) and contained dissolved iron also • Heavy corrosion of joints, gaskets, and gasket seating areas leading to leakage of process water into the condensate • Many subsurface pits with associated crack-like features (blister bursts) were seen on metallographic examination of representative samples of the tubes. Surface pits and general corrosion were also seen on the outside surface of the tubes
DIAGNOSIS	• Subsurface HYDROGEN BLISTERING at depths very close to the outside surface due to corrosion by acidic condensate. Corrosion leads to formation of atomic hydrogen which diffuses into the surface leading to subsurface blistering. This process of diffusion of hydrogen is accelerated by the presence of sulfide which is a poison against evolution of hydrogen as a gas away from the surface. Sulfur/hydrogen sulfide accelerates hydrogen entry into the tube wall • The above process requires a condensed phase and cannot happen on the vapor phase. Hence, the attack has occurred near the condensate outlet region
REMEDY	• The source of acidic sulfide leakage into the upstream feed water in the ISG should be traced and plugged • Change of MOC of the tubes to duplex SS UNS S31803 should be considered

CASE STUDY 52   Textile Bleaching Vessel in a Dyeing Industry

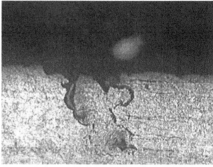

(a) (200X)   (b) (200X)

Case 52   (a) Pitting corrosion initiated on the inside surface of 316 SSTi Vessel. (b) Pitting corrosion followed by cracks initiated on the inside surface of 316 SSTi vessel.

SERVICE	Bleaching of various in-coming textile cloths before further processing, like dyeing, in the vessel at temperatures varying from ambient to 130°C
PROBLEM	Localized pitting corrosion in the bleaching vessel on the wetted inner surfaces, leading to sharp surfaces and rubbing and fiber removal from the cloths, thereby damaging the latter
MATERIAL	Type 316Ti SS
OBSERVATIONS	• Shallow to deep localized corrosion pits on the inner surface giving rough edges • A few torn-off cloths immediately after bleaching process • Both generic chemicals and proprietary chemicals added to the vessel for bleaching and dyeing. The generic chemicals are: sodium chloride, oxalic acid, sodium hydrosulfite, acetic acid, hydrochloric acid, and sulfuric acid. The total chloride content in the bath is about 0.20% including those contributed by proprietary chemicals
DIAGNOSIS	PITTING CORROSION by chloride
REMEDY	• Polish with fine sand paper and follow it up with buffing periodically. Repeat this as often as possible if chloride entry cannot be avoided • Examine using a thin protective Teflon lining on the inside surface of the vessel

CASE STUDY 53   Condenser of an Absorption Chilling Machine in an Air-Conditioning Plant

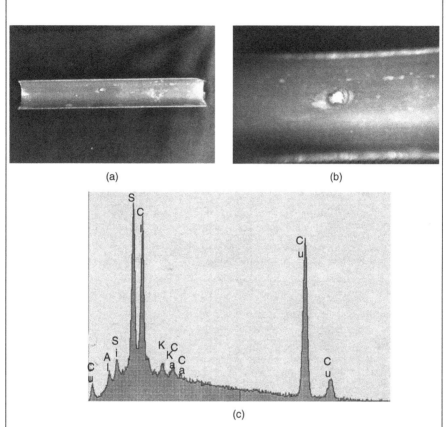

(a)

(b)

(c)

Case 53   (a) Localized deposits on the inside surface of copper tubes. (b) Pitting corrosion under deposits in copper tubes resulting in leakage. (c) EDAX analysis result of corrosion product deposits on copper tubes.

continued

CASE STUDY 53	Continued
SERVICE	Tubes in the absorber and condenser sections of the absorption chilling machine. Tube side: cooling water. Outside: refrigerant
PROBLEM	Leakage within 10 months of operation
MATERIAL	Copper tubes: 19 mm OD × 0.6 mm WT
OBSERVATIONS	• Leakage initiated on the inside surface of the tubes • Uniform corrosion and localized green loose deposit on the inside surface. Below the deposit, fine pinhole pits surrounded by adhering deposit • Raw process water used for cooling purpose. The cooling water analysis showed the following typical values for the mentioned relevant parameters with specification values shown in the brackets: pH 5.4 (7.5–8.0), Sulfate 790 ppm (200 ppm max.), Conductivity 2000 μS/cm (800 max) • The deposits contained mainly copper sulfide, copper chloride, and copper oxide. It also contained 0.4% sulfur. The microbial count was also quite high (much higher than $10^5/cm^2$)
DIAGNOSIS	• GENERAL CORROSION by excess sulfuric acid added to control pH • LOCALIZED PITTING CORROSION by SRB (sulfate reducing bacteria)
REMEDY	• Improve cooling water treatment so that the limits are maintained and ensure that corrosion inhibitors, specifically for copper and proper biocides added to control copper corrosion and to kill SRB

CASE STUDY 54   Petrochemical Complex: Lube Oil Cooler Tubes of Captive Gas Turbine Power Plant

DEZINCIFICATION CORROSION IN BRASS TUBES

SERVICE	Shell and tube coolers for lube oil. Shell side: warm lube oil, cooled from 70 to 55°C. Tube side: cooling water with temperature rise from ambient by 07°C
PROBLEM	Leakage in the tubes in 6–18 months of operation
MATERIAL	Admiralty brass (ASTM B-111/C44300 Type B) 16 mm OD × 1.65 mm WT. Intrinsically finned on the outside
OBSERVATIONS	• Greenish yellow sludge deposit on the tube side • Wide pits along a deep score on the inside surface • Copper coloration on the inside surface at localized places • Free residual chlorine in cooling water frequently crosses the limit of 0.5 ppm max.
DIAGNOSIS	• PLUG TYPE DEZINCIFICATION on the inside surface at both localized places and also at mechanical defects like scores/cracks • Dezincification accelerated by high amount of free residual chlorine in the cooling water
REMEDY	• Periodic inspection, cleaning, and back washing • Control free residual chlorine (FRC) • Introduce corrosion inhibitors like AZOLES which are specifically meant for copper base alloys • As a long-term measure, replace the tubes with cupro-nickel 90/10 or 70/30

CASE STUDY 55   Petrochemical Plant: Plate Heat Exchanger (PHE) Exchanging Heat Between Spent Caustic and Vent Gas in a Cracker Plant

CREVICE CORROSION IN TITANIUM PHE ELEMENTS

SERVICE	PHE: Spent caustic liquor heated from 41 to 110°C on one side by vent gases (Nitrogen, oxygen, and steam) flowing on the other side and getting cooled from 129 to 110°C
PROBLEM	Leakage in about 2½ years
MATERIAL	Titanium plate 0.6 mm thick (ASTM B-265 Gr.1)
OBSERVATIONS	• Heavy deposits in the caustic side to the extent of choking • On cleaning and hydro jetting, leakage detected through punctures below deposits • Deposits could be scraped off with some difficulty. They consisted of some plastic substance • Thickness near the punctures below the deposits is about 0.31 mm
DIAGNOSIS	• CREVICE CORROSION (DIFFERENTIAL AERATION CORROSION) on the caustic side under the polymeric porous deposits thrown by spent caustic
REMEDY	• Polymer addition to process water (added to the caustic) should be reviewed and the polymer changed to one with higher temperature resistance • Periodic cleaning at every available opportunity so that the deposits are washed away and are not allowed accumulating at localized places

CASE STUDY 56 Inorganic Chemical Plant: Distillation Pots

HIGH TEMPERATURE ATTACK LEADING TO MELTING

SERVICE	Direct flame heating (fuel oil burners) and distillation held at 530°C. P2O5 distills off. Sulfur-rich impurities build up as residues. Batch process
PROBLEM	Very short life of pots. 16 mm thick pots last for 16 batches only
MATERIAL	Carbon steel, SA 515 Gr. 70, 16 mm thick
OBSERVATIONS	• At the end of 16 batches, burning leakage holes develop at the bottom • Gradual thickness reduction elsewhere, particularly near the places approaching the very bottom • Inside appearance: grayish rough appearance surrounding the leaky hole with indications of melting on the hole edges • Outside appearance: additional brown rust • Residue is mostly sulfide, lightly water-soluble leading to acidic pH (2–3) • Carbon pick-up on the inside surface (0.59% as against the standard value of 0.20%). Decarburization on the outside surface (carbon level 0.092% as against the standard 0.20%)
DIAGNOSIS	• Extreme high-temperature attack by direct flame accelerated by insulating residues accumulating inside leading to melting
REMEDY	• Redesign the burner position so that only flue gases (not the flame itself) reach the outside of the vessel • Change over to better MOC: 321/310 SS or some other proprietary high temperature grade SS

CASE STUDY 57   Starch Industry: Economizer Tubes of High Pressure Captive Boilers

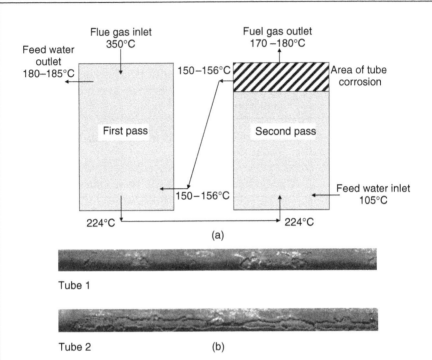

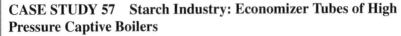

Tube 1

Tube 2   (b)

Case 57   (a) Flow paths of fluids, flue gas and feed water, and temperature profiles in economizer passes and the area of tube corrosion. (b) Corrosion patches on the outside surface of economizer tubes-1. (c) Corrosion patches on the outside surface of economizer tubes-2.

SERVICE	Economizer tubes of captive steam boiler
PROBLEM	Corrosion on the outside surface: flue gas side
MATERIAL	Boiler quality carbon steel
OBSERVATIONS	• Mildly corroded longitudinal localized areas on the outside surface of the economizer tubes • The above attack is only on tubes well away from the soot blowers
DIAGNOSIS	DEW POINT CORROSION during idle period $(S^- + H_2O + O_2 \rightarrow H_2SO_3 \rightarrow H_2SO_4)$
REMEDY	• Modify soot blowing system to make it more efficient • Adopt alkaline wash immediately prior to shutdowns

CASE STUDY 58   Rubber Chemical Plant – Crump Slurry Tank

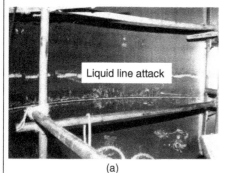

(a)

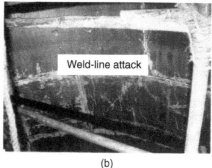

(b)

(c)

Case 58   (a) Corrosion of Monel vessel by acidic chloride water at liquid vapor interface. (b) Corrosion of Monel vessel by acidic chloride water on weld line and nearby places. (c) Corrosion of Monel vessel by acidic chloride water at weld heat tint zones.

continued

CASE STUDY 58   Continued

SERVICE	Washing rubber crumbs with live steam to remove chemicals associated with crumbs. Recycle water is added to the tank pH: 2–4, chloride: large quantity, temperature: 91–98°C, pressure: 0.2 kg/cm^{2}g
PROBLEM	After 5 years of operation, with periodic and frequent shutdowns, unusual corrosion/erosion patches noticed on the inside surface
MATERIAL	Shell: Monel 400 (ASME SB 127). 5 mm thick plate, 4.5 M Ø, 8.5 M height Impeller blades: 304 L SS
OBSERVATIONS	• Several corrosion indications throughout the inside surface, more near weld areas, near water-mark lines and on heat-tinted locations corresponding to outside welds and tag welds • Impeller blades had many branched cracks
DIAGNOSIS	• Corrosion by acidic wash water in the presence of dissolved oxygen during shutdown periods in which the water was not drained. (Monel is resistant to acidic chloride solution, only in the absence of dissolved oxygen) • STRESS CORROSION CRACKING by chloride of impeller blades of 304 L SS
REMEDY	• Drain wash water during shutdown, wash the vessel with fresh water, and keep the vessel dry • Replace 304L SS with 316 SS/duplex SS for the impeller blades

CASE STUDY 59 Petrochemical Plant: Gas Cracker Unit, Dilute Steam Kettle Re-boiler	
ACIDIC IMPURITY CORROSION AND UNDER-DEPOSIT CORROSION	
SERVICE	U-tube kettle reboiler producing dilute steam necessary for cracker. Shell side: process water and boiler feed water at 7.5 bar, 180°C Tube side: steam at 12.0 bar, at 250°C
PROBLEM	Leaks in about 16% of the tubes within 7 months of operation
MATERIAL	Carbon Steel Seamless Tubes, A-179, 19.05 mm OD × 2.3 mm WT, 6 M strength length, 1110 tubes
OBSERVATIONS	• Severe corrosion with non-adherent brownish deposit on the OD surface of tubes rows on the top portion (tube-side entry) • Excessive wall thinning, puncture, and roughening of the tube surface. • Adherent grayish carbonaceous deposit with localized pitting and puncture on a few bottom row tubes • Acidic deposit, lower pH in top deposits than bottom deposits, carbon content in the bottom deposits high • Sulfate is high in the top deposit, same with chlorides
DIAGNOSIS	• Top rows: ACIDIC CORROSION by H_2SO_4 ingress to boiler feed water from water treatment plant • Bottom rows: UNDER-DEPOSIT-CORROSION due to carbonaceous deposit from the process stream
REMEDY	• Review the boiler feed water treatment and take action so that boiler feed water is free from acid • Periodically clean with high pressure water jet the outside surfaces of the tubes to remove loose deposits

CASE STUDY 60   Textile Dyeing Unit: Jet Dyeing Machine Shell

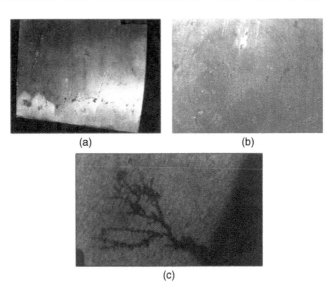

(a)   (b)

(c)

Case 60   (a) Localized pits and cracks as seen on the outside surface of 316Ti SS jet dyeing machine shell. (b) Localized cracks as seen on the outside surface of 316Ti SS jet dyeing machine shell. (c) Cross sectional micrograph of chloride stress corrosion cracking initiating below corrosion pits in 316Ti SS(50X).

SERVICE	Machine is used for jet dyeing of textile fabrics
	Dye solution: An acidic chloride solution neutralized to pH 5.5 to 6.5 by caustic addition and processed at 120–140°C
PROBLEM	Cracking within 8 months resulting in leakage
MATERIAL	Type 316 Ti SS, 3 mm thick, roll formed and welded
OBSERVATIONS	• Multiple cracking, initiated at different places on the outside surface and propagated with branching towards the inside surface resulting in leakage
DIAGNOSIS	• CHLORIDE STRESS CORROSION CRACKING on the external surface due to the leakage of dye liquor through flange joints
REMEDY	• Tighten flange joints systematically so that there is no leakage from gasket areas • Use duplex SS if leakages cannot be arrested

CASE STUDY 61   Oil Refinery: 12 Inch Dia. Overhead Pipeline

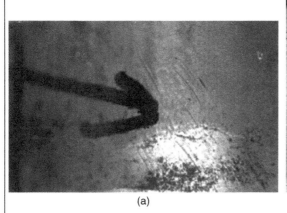 

(a)   (b)

Case 61   (a) Cracks as seen on the external, outside, surface of overhead pipeline of 304SS in an oil refinery. (b) External stress corrosion cracking (ESCC) due to chloride corrosion under insulation (CUI) on 304 SS pipe in an oil refinery 500X.

SERVICE	12″ NB 40 Sch. Pipe (4.36 mm WT). Pipeline from regenerator to overhead condenser transferring H_2S-rich acid gas (80% H_2S) at a pressure of 1 kg/cm^2 and at a temperature of 110°C
PROBLEM	After 10 years of service, leaking occurred during hydro-testing
MATERIAL	304 SS, welded pipe
OBSERVATIONS	• Cracks initiated on localized pits on the outside surface below a simple support • Fine multiple cracks
DIAGNOSIS	• External chloride stress corrosion cracking due to sudden wetting of insulation material containing 7 ppm chloride
REMEDY	• Ensure that insulations are free from chloride and that they are kept dry throughout by proper overlapping of external protective cladding

CASE STUDY 62   Fertilizer Plant: CO_2 Compressor Inter-stage Cooler

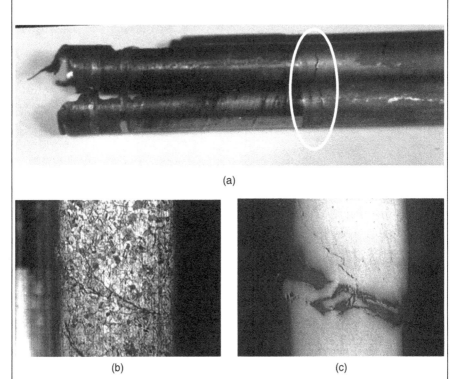

(a)

(b)   (c)

Case 62   (a) Transverse cracks as seen on the outside surface of 304L SS tubes little away from the expanded ends. (b) Chloride stress corrosion cracking initiated on the inside surface of compressor inter-stage cooler tubes of 304L SS(50X). (c) A fully grown stress corrosion crack, from inside to outside, of compressor interstage cooler tube of 304L SS(50X).

SERVICE	Horizontal shell and tube cooler
	Tube side: hot CO_2 gas entering at the top half from second stage compressor outlet at a pressure of about $20\,kg/cm^2$ gets cooled from 193°C at the inlet to 44°C at the outlet
	Shell side: cooling water entering at ambient temperature at the inlet absorbs heat from the gas and gets heated to about 44°C

CASE STUDY 62   Continued

PROBLEM	Leakage of tubes within 9 months of service
	pH of cooling water has gone down to 4.5 from 7.5
MATERIAL	Type 304L SS Seamless tubes as per ASTM A-213, 19.05 mm OD × 1.24 mm WT × 5500 mm long and 431 Nos. Two pass design
OBSERVATIONS	• 56 tubes on the outer rows of top half (tube-side entry) have leaked
	• Sharp transverse cracks on positions just after the tube sheet inner edge on the gas inlet side
	• Branched transgranular cracks initiated on the inside surface of the tubes (gas side)
	• The in-coming compressed gas contains moisture from the spray cooler upstream
	• Instead of the specified DM water for spray cooling purpose, cooling water with 250–350 ppm chloride has been used frequently whenever DM water shortage arises
	• There has been frequent noticeable vibrations in the cooler
DIAGNOSIS	• Chloride stress corrosion cracking initiated on the inside surface of the tubes on stressed portions near the tube-expansion ends at the high temperature regions
REMEDY	• Use only DM water for spraying purpose upstream
	• Reduce vibration levels
	• Consider tube MOC change to duplex SS UNS S31803

CASE STUDY 63   Oil Refinery: Flash Crude Heater Shell Cover Drain Nozzle

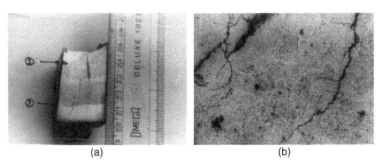

(a)   (b)

Case 63   (a) Cross section of crack across nozzle, weld and shell. (b) Caustic intergranular stress corrosion cracking in carbon steel shell plate (100X).

SERVICE	Horizontal shell and tube heat exchanger, shell side: flash crude heated from 200 to 260°C at 24 kg/cm^2 pressure, tube side: heavy gas oil cooling from 308 to 260°C
PROBLEM	Within 3 ½ years of operation, leakage noted in the shell bottom drain nozzle area
MATERIAL	Carbon steel. Shell: SA 516 Gr. 60, 28 mm thick, drain nozzle: seamless pipe as per SA 106 Gr. B 2″ NB Sch 160
OBSERVATIONS	• Radial cracks from the nozzle into the shell through the weld and towards the outside surface resulting in leaks • Multiple crack initiation sites with intergranular and branched propagation towards the shell outside surface • Hardness values in VHN: nozzle 152, weld 161 and shell 165
DIAGNOSIS	• Caustic stress corrosion cracking. Caustic alkali is added to the crude to neutralize the acidic salts present in the raw crude. Concentration of the caustic at the bottom drain area over the years
REMEDY	• Use the drain nozzle periodically so that caustic does not get concentrated at the bottom • Ensure that not very excessive caustic is added to the raw crude • Do local stress relief heat treatment after replacement of the nozzle and repair of the cracks

CASE STUDY 64   Oil Refinery: Light Cycle Oil Steam Generator

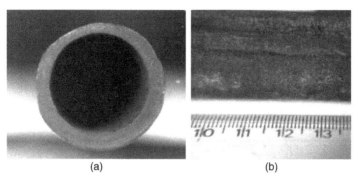

(a)	(b)

Case 64   (a) Cross section of a steam generator tube showing thickness reduction on the outside surface. (b) Close view of the corrosive attack on the outside surface.

SERVICE	Horizontal shell and tube heat exchanger generating steam. Tube side: light cycle oil entering at 293°C and leaving at 173°C and at 10.4 kg/cm^2 pressure. Shell side: boiler feed water getting converted to steam at 160°C at a pressure of 5.2 kg/cm^2
PROBLEM	Tube leakage near the expanded portion in the end, near the hot oil entry
MATERIAL	Carbon steel. Seamless tubes as per ASTM A-179, size 25.4 mm OD × 2.80 mm WT × 16 feet long and 260 Nos.
OBSERVATIONS	• Four tubes from the top most row of the bundle, close to the hot oil entry, leaked • 13 second row tubes showed grooving, pitting, and general roughness on the upper half of the outside surface to a length of about 1–2 m from the hot oil entry end • Chemical analysis showed pH of feed water to be 9.8 as against the specified range of 7–8 and that of blow-down water to be highly alkaline, 13.8
DIAGNOSIS	• CAUSTIC GOUGING (erosion corrosion by concentrated caustic) at places where there is heavy flow of fluid carrying the caustic, in this case steam with unconverted water strong in alkali concentration
REMEDY	• Maintain feed water and blow down water chemistry within limits

CASE STUDY 65   Organic Chemical Plant: High Pressure Autoclave in R&D Laboratory

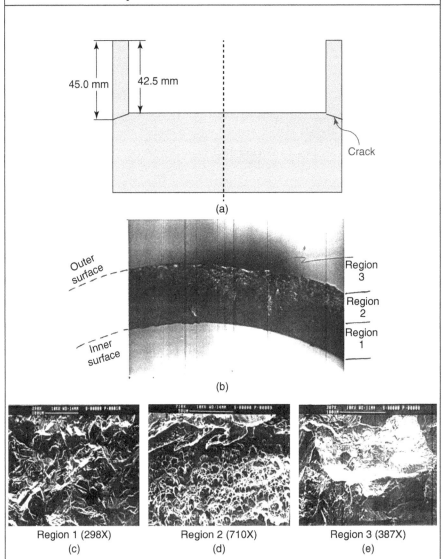

(a)

(b)

Region 1 (298X)   Region 2 (710X)   Region 3 (387X)
(c)   (d)   (e)

Case 65   (a) Elevation of a section of an autoclave showing the leakage position. (b) Cross sectional photograph of the crack surface showing three regions of crack propagation. (c) SEM fractograph of region 1. Intergranular initiation (298X). (d) SEM fractograph of region 2. Mechanical propagation (710X). (e) SEM fractograph of region 3, ductile tearing (387X).

CASE STUDY 65	Continued
SERVICE	Laboratory autoclave processing batch-wise operations involving different chemicals at different times under high pressure and at elevated temperatures
PROBLEM	Sudden leakage through a circumferential crack after 5 years of service
MATERIAL	Type 316 SS, 122 mm ID, 9.5 mm shell wall thickness, 35 mm base thickness and 255 mm height
OBSERVATIONS	• Two circumferential cracks somewhere on the lower half of the shell on the same horizontal plane • The plane is that of the inside bottom surface • The fracture surface, after opening the crack, on SEM examination, showed three regions: inner region: gradual corrosion with intergranular penetration, middle region: mechanical propagation, and outer region: ductile final tearing
DIAGNOSIS	• CREVICE CORROSION at the unsmoothened (absence of smooth radius) sharp inside edge between the bottom surface and the shell wall. Reaction acids under high pressure penetrate the crevice and lead to corrosion at the crevice. Wash water between batches at atmospheric pressure does not remove the acids and crevice corrosion continues to occur. After reaching a certain depth, under high operating pressure, crack develops from the crevice leading to through and through cracking followed by tearing
REMEDY	• Provide smooth large radius at the inside edge

CASE STUDY 66   Fertilizer Industry: Captive Power Plant: Economizer Tube

(a)   (b)

Case 66   (a) Leaking hole on a boiler economizer tube and the erosive effect on the neighboring tube. (b) Close up view of oxygen pitting corrosion on the inside surface of a boiler economizer tube.

SERVICE	Economizer tube of captive power plant boiler. Feed water on the inside gets heated from 213 to 286°C at 120 kg/cm^2 pressure
PROBLEM	Leakage after 13 years of service
MATERIAL	Boiler quality carbon steel as per BS-3059. 48.3 mm OD × 3.2 mm WT
OBSERVATIONS	• Many fine pits on the inside surface • One such pit enlarged and deepened to leakage • The leaking high pressure water impinged on the neighboring tube resulting in impingement attack
DIAGNOSIS	• PITTING CORROSION by dissolved oxygen in the feed water. The DO content was not monitored regularly. Thermal deaerators and certain batches of added hydrazine were not effective suddenly
REMEDY	• Ensure that dissolved oxygen is maintained at levels less than 5 ppb in the deaerated feed water to the economizer

CASE STUDY 67   Inorganic Chemicals Plant: Reactor Shell
CORROSION OF STAINLESS STEEL BY REDUCING ACID

SERVICE	Sulfuric acid neutralization tank in a silica crystallizer reactor
PROBLEM	Excessive general corrosion on the inside surface and on the interior baffles at certain localized places only
MATERIAL	Type 304 SS
OBSERVATIONS	• Baffle plates have become wafer thin and the welds were eaten away • Shell wall attacked to a great depth with general corrosion and wall thinning • Thick silica deposit with heavy metal impurities • PVDF acid distribution pipe at the top has distribution holes facing the shell wall instead of facing vertically downwards at the liquid below
DIAGNOSIS	• General corrosion by sulfuric acid due to the splashing of the acid directly on the shell wall instead of on the alkaline liquid reactor mass below
REMEDY	• Reposition the top distribution pipe so that the holes have exact vertical axis • Cover the affected areas with 316L SS patch plates

CASE STUDY 68   Reactor for an Organic Chemical Plant

Case 68   (a) Corrosion pits on a baffle support bracket and nearby areas on the shell. (b) Corrosion pits on the shell above the weld line.

SERVICE	The reactor is a part of an organic chemical plant which generates and uses hydrochloric acid gas
PROBLEM	Reactor was obtained fresh and was not used for 6 years. First few batch operations gave very high concentration of heavy metal content in the product. Reactor was stopped and stored for a period of 3 months with nitrogen atmosphere inside. On opening; corrosion marks (rusting/pitting) were seen
MATERIAL	316 SS
OBSERVATIONS	• Heavy pitting marks on the inside surface • Rust patches, with deep pits at the center of each patch, on crevice areas such as baffle support brackets
DIAGNOSIS	• ATMOSPHERIC HCl CORROSION during the 6 year storage period • Further rusting and pitting marks during nitrogen blanketing (free of oxygen)
REMEDY	• Passivate with dilute nitric acid containing sodium dichromate at about 70°C, wash, dry, and use

CASE STUDY 69   Fertilizer Unit. Ammonia Plant. Primary Waste Heat Boiler

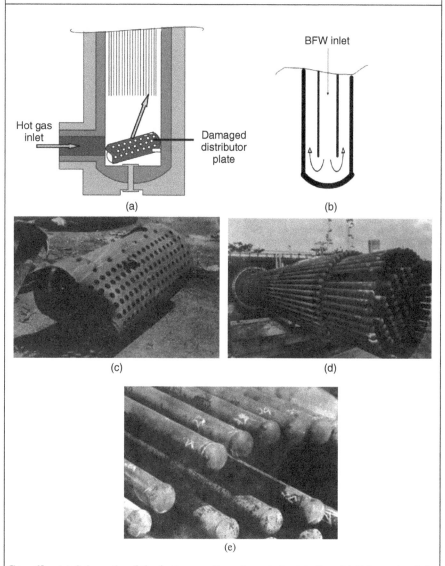

(a)

(b)

(c)

(d)

(e)

Case 69   (a) Schematic of the bottom portion of waste heat boiler. (b) Schematic of the concentric steam generator tube assembly with the bottom closure cap. (c) Photograph of damaged gas distributor. (d) Affected steam generator tube bundle with closure end caps at the bottom ends. (e) Steam generator tubes with blisters on the closure end caps.

continued

CASE STUDY 69	Continued
SERVICE	Concentric tube primary waste heat boiler, shell side: process gas: with inlet temperature 940°C and outlet temperature 411°C at a pressure of 31 kg/cm²g, tube side: boiler feed water and steam at 315°C and at a pressure of 105 kg/cm²
PROBLEM	• Breakage of bottom perforated distributor plate • "Blisters" in about 10% of 255 hanging tubes on the closure caps at the bottom of the outer tube
MATERIAL	Bottom plate: 310 SS/304 SS Tube: SA 209/T1 C-Mo alloy steel Cap: SA-182, C-0.5 Mo alloy steel forged or rolled
OBSERVATIONS	• Blisters on bottom caps of almost half of the tubes, on the half portion of the bundle, opposite the gas inlet nozzle • The blistered caps were very low in thickness (only one third) below the blisters • Pitting attack on the outside surface of the inner tube of steam generator bundle • Hardness values: blister areas: 124 VHN, non-blister areas: 182 VHN • Microstructure: decarburization, large grain size and plastic deformation
DIAGNOSIS	• Wrong choice of MOC for the bottom distributor plate has resulted in premature breakage. This has led to non-uniform heating of the outer tubes. Almost half of the tubes were over-heated resulting in plastic deformation, wall thinning, and blisters
REMEDY	• Change the distribution plate to a better high temperature grade SS having much higher resistance to high temperature phenomena than that of Type 310 SS

CASE STUDY 70   Pulp and Paper Plant. TL Vertical Screen Inlet Line of the Paper Section

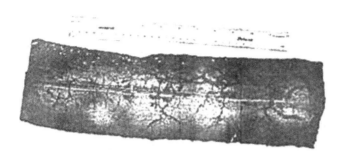

Case 70   Cracks as seen on the outside surface of \ pipeline along and across the weld line.

SERVICE	Pipeline from the outlet of primary centricleaners to the inlet of the vertical screen in the paper mill 400 mm OD × 4.00 mm WT × 4.7 m length
PROBLEM	Sudden burst after 15 years of service giving rise to huge loss of stock pulp
MATERIAL	Type 316 SS, fabricated pipe from plate, as per ASTM A 312 specification
OBSERVATIONS	• Rust patches associated with cracks seen on OD surface. No problem on the inside surface • Cracks on parent metal, weld metal, and heat-affected zone • Survey of chemical analysis of atmospheric moisture showed chloride level maximum 4 ppm while the normal range is about 0.38 ppm • Cross-sectional microscopic examination showed sensitization throughout the thickness of the pipe and corrosive attack from the outside surface towards the inside surface through the grain boundaries • Chemical analysis of the pipe material showed it to be 304 SS instead of 316 SS
DIAGNOSIS	INTERGRANULAR CORROSION by the outside wet atmosphere over the years of operation due to the use of sensitized (not properly solution annealed) 304 SS
REMEDY	Replace the pipe with properly solution-annealed 316L SS

CASE STUDY 71 Fertilizer Plant. Underground Sections of Cooling Water and Fire Hydrant Water Pipe Lines

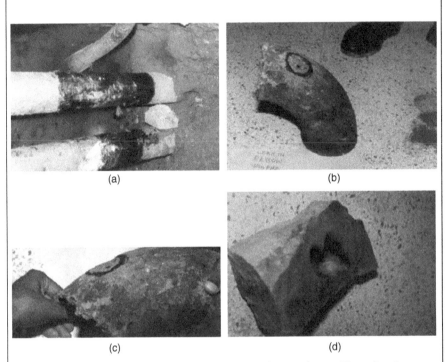

(a)   (b)

(c)   (d)

Case 71 (a) Leaky spot and surrounding corroded regions on the outside surface in contact with under-ground soil. (b) Close-up view of a leaky hole in one of the elbows on the outside surface. (c) Closer view of the leaky spot on the outside surface of the elbow showing that the attack is from outside surface. (d) Typical boulder having sharp edges and corners.

SERVICE	Underground pipelines of fire hydrant water systems and cooling water systems
PROBLEM	A few leaks within 1 ½ years of installation
MATERIAL	Carbon steel (API 5L.Gr.B) 2″ to 24″ NB. WT from 3 to 10 mm. Outside surface protection by coal tar primer, coal tar enamel coating, and fiber glass tissue inner wrapping and fiber glass felt outer wrapping

CASE STUDY 71	**Continued**
OBSERVATIONS	• Leaks have occurred only in those places where coating was damaged • In many places coating has lost its adherence to the pipe because of a high water table • All the leaks have occurred only on top half: from 11 o'clock position to 1 o'clock position • Initiated as pitting corrosion on the external, soil side, surface • Surrounding boulders had sharp edges and corners
DIAGNOSIS	Leaks because of external corrosion. The latter has occurred essentially because of coating damage. This in turn occurred as sharp edged boulders strike the coated pipe while initial laying of pipes and filling of trenches
REMEDY	Weld repair the leaky holes, rectify the coating damage, and install cathodic protection system

CASE STUDY 72   Thermal Power Plant. Condenser Cooling Sea Water In-take Line. Butter Fly Valve

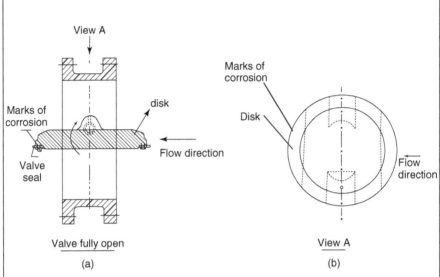

Case 72   (a) Sketch of the condenser inlet valve. Longitudinal section with valve fully open. (b) Sketch of the cross section of the valve disc indicating the damage locations.

SERVICE	2200 mm diameter butter fly valve in sea water in-take line
PROBLEM	The valve disc showed "corrosion marks" after 1 year of operation
MATERIAL	Ni-resist cast iron
OBSERVATIONS	• Corrosion marks at different points on the edge of the disc on the side facing the down-stream debris filter
DIAGNOSIS	• Local erosion due to turbulence at the downstream proximity of debris filter. Local erosion only at the leading edge of the valve disc
REMEDY	• Provide a hard wear-resistant coating on the affected areas and other areas likely to be affected

CASE STUDY 73   Petrochemical Plant. Pressure Transmitter Sensors

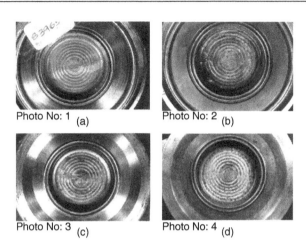

Photo No: 1 (a)   Photo No: 2 (b)

Photo No: 3 (c)   Photo No: 4 (d)

Case 73   (a) Sensor diaphragm of pressure transmitter with corrosion marks. Water line attack. (b) Sensor diaphragm of pressure transmitter with corrosion marks. Uniform and pitting corrosion. (c) Sensor diaphragm of pressure transmitter with corrosion marks. Water line attack. (d) Sensor diaphragm of pressure transmitter with corrosion marks. Uniform and pitting corrosion.

SERVICE	Sensor diaphragms of pressure transmitters in different process media, organic process fluids, benzene and phenol with and without water
PROBLEM	Failure of the diaphragms within a short time
MATERIAL	Type 316L SS. Very thin foil, thickness a small fraction of a mm
OBSERVATIONS	• Sponginess and black spots on the exposed diaphragm • Loss of sensing activity of the diaphragm • Heavy brownish rusty deposit with horizontal water mark on the exposed surface of the diaphragm
DIAGNOSIS	• Stagnation of the process liquid leads to sediments, deposits, and pitting followed by leakage and failure of the sensing ability of the diaphragm • Partial stagnation leads to water mark corrosion on the liquid vapor interface
REMEDY	• Provide upstream filters so that sediments do not come in contact with the sensor diaphragm • Ensure that stagnation, both full and part, does not occur

CASE STUDY 74   Organic Chemicals Plant: Coolers and Condensors

(a)   (b)

(c)

Case 74   (a) General corrosion on the inside surface by cooling water followed by deposits and under deposit corrosion. (b) General corrosion on the inside surface by cooling water followed by deposits and under deposit corrosion. (c) Corrosion by cooling water on inside surface. The leaking pit as seen on the outside surface.

SERVICE	Shell and tube coolers and condensers. Shell side: mixture of amines, ammonia, alcohol, and water. Inlet temperatures vary from 55 to 175°C in different services. Tube side: cooling water
PROBLEM	Tube leakages within a short period of less than 1 year
MATERIAL	Carbon steel seamless tubes as per ASTM A-179 specification. Size: 25.4 mm OD × 14G × 3660 mm long

CASE STUDY 74 Continued	
OBSERVATIONS	• Outside surfaces of the tubes: no significant noticeable corrosion • Inside surface: adherent brownish gray deposit scale with localized heavy build up and pitting corrosion under the deposit leading to leakages • Collection of greenish wet sludge (biomass) on the tube-side outlet channel box • The inside deposit scale contains calcium, magnesium, iron, and biomass
DIAGNOSIS	• General and localized corrosion by the cooling water leading to surface roughness and corrosion product porous deposit. Porosity helps in water stagnation under the deposit. Biomass adds to the deposit. The water-borne calcium and magnesium salts also add to the deposit • Under-deposit-corrosion pitting on the inside surface below the above mentioned deposits, more predominant on higher temperature zones • Thermal cycling between two product campaigns aggravates cracking of adherent cooling water side scales and accelerates pitting corrosion • Ammonia is a good nutrient for micro-organism and hence leads to high growth of green biomass (algae and bacteria) which in turn promotes under-deposit-pitting-corrosion
REMEDY	The following modifications/improvements should be incorporated in the chemical treatment of cooling water 1. Passivation prior to commissioning by dosing high concentrations of anodic inhibitors 2. Good organic antiscalants should be added to the cooling water treatment package 3. Good effective micro biocides must be added to the cooling water as a part of the overall chemical treatment to kill bacteria thriving in cooling water

CASE STUDY 75   Organic Chemicals Plant: Alcohol Vaporizer

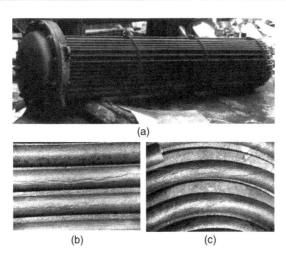

(a)

(b)   (c)

Case 75   (a) Alcohol vaporizer bundle. (b) Alcohol vaporizer bundle. Erosion grooves on straight portions of tubes. (c) Alcohol vaporizer bundle. Erosion grooves on U-bent portions of tubes.

SERVICE	Kettle design horizontal reboiler. Shell side: mostly alcohol, balance water at 155°C and 17 kg/cm² pressure. A gas mixture of nitrogen and hydrogen also enters the shell side. Tube side: steam at 220°C and 16 kg/cm² pressure
PROBLEM	Tube leakage with service life of 3–4 years of service
MATERIAL	U-bent carbon steel seamless tubes as per ASTM A-179 specification. 25.4 mm OD × 3.4 mm WT × 2500 mm straight length, 106 Nos.
OBSERVATIONS	• Deep grooves and valleys, both longitudinal and transverse, on the outside surface of both straight and u-bent portions of the tubes • Leaky tubes were mostly from interior positions
DIAGNOSIS	Gradual erosion of the top half of the tubes exposed to flowing mixtures of gases and vapors of alcohol and water
REMEDY	• Provide two vapor outlets instead of one • Ensure the full bundle is always in the liquid phase when evaporation takes place • Consider changing MOC to duplex stainless steel UNS S31803, a material of good erosion resistance

CASE STUDY 76   Organic Chemicals Plant. Thermowells

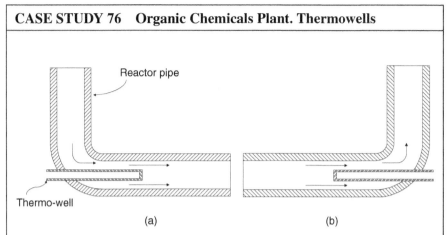

Reactor pipe

Thermo-well

(a)   (b)

Case 76   (a) Sketch of the orientation of un-affected thermo-well with respect to flow of process fluid. (b) Sketch of the orientation of affected thermo-well with respect to flow of process fluid.

SERVICE	Thermowells inserted horizontally into a coiled reactor of several turns processing organic chlorides
PROBLEM	Leakage in the closed end of the thermowells
MATERIAL	316 SS. Hollow bars. 30 mm OD × 10 mm ID × 200 mm length, one end not machined out
OBSERVATIONS	• The closed ends showed many sharp fine needle-like straight holes penetrating through and through the whole thickness • Longitudinal grooves seen on the outside horizontal surfaces of the thermowells • Only certain thermowells showed leakage, not all • Those with closed ends facing the moving fluids head-on have failed while those on which the moving fluid slides past the horizontal surface from behind have not failed
DIAGNOSIS	EROSION CORROSION due to direct impingement of the flowing fluid head-on on the closed end surface of the thermo wells
REMEDY	Re-design the location of the thermowells so that gases slide past the horizontally placed thermowells over the outside surface and not strike them head-on

CASE STUDY 77   Fertilizer Unit. Ammonia Plant. Gas to Gas Heat Exchanger

GENERAL CORROSION BY CO_2 ACIDIFIED SOLUTION

SERVICE	Shell and tube heat exchanger. Tube side: hot synthesis gas entering at 129°C and 52 kg pressure gives away the heat and gets cooled down. Shell side: methanator feed cold gas entering 70°C and 29 kg pressure absorbs heat and gets heated up
PROBLEM	Tube leak after only 5 months of operation
MATERIAL	Carbon steel seamless tubes as per ASTM A-179 specification, 2.1 mm wall thickness
OBSERVATIONS	• Heavy uniform corrosion with leakage holes on the outside surface of the tubes • The attack is only up to length of about 30″ from the shell side inlet nozzle position • Carryover of CO_2-rich Benfield solution with methanator feed gas has been substantial • The approximate corrosion rate of the leaked tube-walls near the leaks is about 6 mm/year, an unacceptably high corrosion rate
DIAGNOSIS	• Uniform corrosion by the corrosive acidic CO_2-rich Benfield solution coming along with the shell side feed gas. The attack is only up to a length from the shell side feed nozzle, beyond which the solution vaporizes. Hence, no corrosion beyond 30 inches
REMEDY	• Reduce Benfield solution carry-over by improving the efficiency of the demister pads upstream • Consider using 304L SS as an alternate material of construction

CASE STUDY 78   Oil Refinery. Sulfolane Recovery Column Reboiler

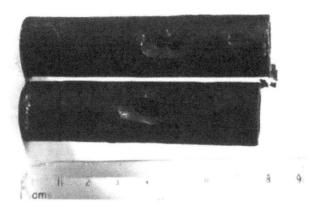

Case 78   Corrosion under sludge by sulfuric acid on carbon steel tubes.

SERVICE	U tube kettle type reboiler. Shell side: sulfolane water mixture boils at about 200 to 220°C. Tube side: hot oil entering at 220°C. Hot oil enters at the bottom half of the tubes and leaves at the top half
PROBLEM	Leaking of tubes within a short period
MATERIAL	Carbon steel seamless tubes as per ASTM A-179 specification. 20 mm OD × 2.0 mm WT × 2250 mm straight length
OBSERVATIONS	• Dark brown sludge coating on the outside surface of the tubes • Below the sludge, uniform corrosion with leaky holes underneath • Leaked tubes are from the bottom two rows only, hot oil entry side • pH of the fluid on the shell side was 4.5
DIAGNOSIS	General corrosion by sulfuric acid on the outside surface of the tubes. Sulfolane decomposes at about 220°C to sulfur dioxide which oxidizes and dissolves in the water present producing sulfuric acid
REMEDY	• Reduce operating temperature if possible and permissible to less than 220°C • Consider high alloy stainless steels or nickel base alloys as a better MOC for the tubes

CASE STUDY 79   Oil Refinery. Hydrocarbon vapor-liquid heat exchanger

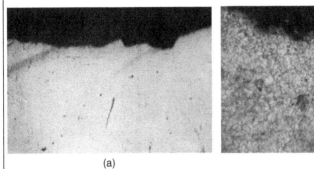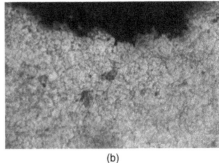

(a)   (b)

Case 79   (a) Corrosion by polythionic acid on alloy steel. As polished outside surface 50X.
(b) Corrosion by polythionic acid on alloy steel. As etched outside surface 225X.

SERVICE	Shell and tube heat exchanger. Shell side: vapors of the cracked crude entering at 386°C and leaving at 280°C. Tube side: long residue liquid entering at 127°C and leaving at 232°C
PROBLEM	Leakage when recommissioning after 6 months of shutdown which followed 27 years of eventless service
MATERIAL	7 Cr-0.5 Mo alloy steel as per T7-A213 specification. 1.00″ OD × 2.11 mm WT × 14′ long
OBSERVATIONS	• Heavy scaling and general corrosion on the outside surface with localized deep corrosion at several places • Chemical analysis of the scale indicated high quantity of iron and substantial quantity of sulfur
DIAGNOSIS	• Corrosion by polythionic acid during the 6 month long shutdown period. Crude contains 2.65% sulfur. Cr-Mo alloy steels would resist high temperature attack by sulfur gases but would corrode at ambient temperatures by polythionic acid which forms at shutdown conditions only
REMEDY	• Keep the tubes in dry inert condition during long shutdown periods preceded by an alkaline wash right at the start of the shutdown operations

CASE STUDY 80   Chlor-Alkali Plant, Stainless Steel Laboratory Reactor

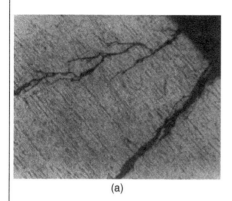
(a)

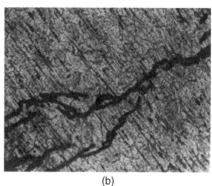
(b)

Case 80   (a) Initiation of caustic stress corrosion cracking in 316SS autoclave shell (50X). (b) Propagation of caustic stress corrosion cracking in 316SS autoclave shell, 100X.

SERVICE	Pilot plant/laboratory reactor producing organic acid from castor oil and fused caustic alkali at 280°C, cooled by direct water addition to 80°C. Batch process. 8 hours per batch
PROBLEM	Cracks on the reactor after only eight batches, within 1 month of commissioning
MATERIAL	Type 316 SS/304 SS
OBSERVATIONS	• Cracks on the inside surface near all welds and at places corresponding to outside welds (limpet coils) • Cracks on thermowell outside surface • Branched transgranular crack • Earlier, laboratory 304 SS reactor without weld which lasted for 50 batches with direct heating showed uniform corrosion at a high unacceptable rate of 3.9 mm/year
DIAGNOSIS	CAUSTIC STRESS CORROSION CRACKING
REMEDY	• Stress relieve the reactor after fabrication and before commissioning • Use pure Nickel or Inconel 600 as a better MOC

INDEX

Corrosion Failures: Theory, Case Studies, and Solutions, First Edition. K. Elayaperumal and V. S. Raja.
© 2015 John Wiley & Sons, Inc. Published 2015 by John Wiley & Sons, Inc.